Second Edition

Laboratory Experiments for Organic & Biochemistry

FREDERICK BETTELHEIM
JOSEPH LANDESBERG
Adelphi University

Saunders Golden Sunburst Series

SAUNDERS COLLEGE PUBLISHING
Harcourt Brace College Publishers

Fort Worth • Philadelphia • San Diego • New York • Orlando • Austin
San Antonio • Toronto • Montreal • London • Sydney • Tokyo

Printed in the United States of America.

Bettelheim/Landesberg: Laboratory Experiments For Organic and Biochemistry, 2/e, to accompany INTRODUCTION TO ORGANIC AND BIOCHEMISTRY, 2/e

ISBN 0-03-001268-7

456 082 987654321

This book is dedicated to our wives:

Vera S. Bettelheim and Lucy Landesberg

whose help, understanding and patience enabled us to write this book.

PREFACE

In writing this laboratory manual three goals were kept constantly in mind: (a) the experiments should illustrate the concepts learned in the lecture; (b) it should be clearly and concisely written so that students will understand easily the task at hand and will be able to perform the experiments in a 2 to 2 1/2 hours lab period; (c) to select experiments of which 85% are optimized conventional experiments and only 15% are of completely new design. It did not escape our attention that in adopting this text of Laboratory Experiments, the instructor must pay attention to the cost. Thus, new and novel experiments, beyond the traditional ones, can be introduced only gradually, especially if this requires the purchase of new equipment. All experiments in this manual require only inexpensive equipment, if any. A few spectrophotometers and possibly a few pH meters (not mandatory) are the necessary instruments.

Twenty four experiments in this book will provide suitable choice for the instructor who usually selects about 12 experiments for a one semester course. The following are the principal features of this book:

1. Ten experiments illustrate the principles of organic chemistry and 14 of biochemistry. Whenever it was feasible the health aspects of the experiments were emphasized.
2. Each experiment has background information in which all the relevant chemical principles and applications pertaining to the experiment are reviewed. This is not a repetition of the textbook material, but we adapted it and extended the concepts to the tasks to be performed.
3. The procedure part provides a step by step description of the experiments. Clarity of this section is of utmost importance and special emphasis was placed in writing to accomplish it. Whenever one deals with dangerous chemicals, strong acids or bases, the **Caution!** sign alerts the students to perform the task with utmost care. We eliminated all carcinogenic substances

from the experiments. Some chemicals are poisons at high concentrations. When we had to use some of these substances, for example mercury, the amounts and the time of exposure were minimized so that students are not exposed to health hazards.

4. Pre-lab questions are provided with each experiment to familiarize students with the concepts and procedures before they start the experiment. By requiring students to answer these questions and by grading their answers, we accomplish the goal of preparing the students for the experiments.
5. In the Report Sheet we not only ask for the raw data of the observations, but we also request some interpretation of the results.
6. The Post-lab questions are designed so that the student should be able to reflect upon the results and to relate their significance.
7. At the end of the book we provide the Stockroom Personnel with detailed instructions on how to prepare the solutions and other chemicals for the experiments. We also give detailed instruction how much material is needed for a class of 25 students.

An Instructor's Manual accompanies this book solely for the use of the instructor. It helps in the grading process by providing possible ranges for the experimental results. In addition, it alerts the instructor to the difficulties one may encounter in each experiment.

We hope that you will find our book of Laboratory Experiments helpful in instructing your students in chemistry and that your students will like it also.

Frederick A. Bettelheim
Joseph Landesberg

Acknowledgments

These experiments have been used by our colleagues over the years and their criticism and expertise were instrumental in refinement of the experiments. We thank Robert Halliday, Cathy Ireland, Mahadeva Kumbar, Jerry March, Donald Opalecky, Reuben Rudman, Charles Shopsis, Kevin Terrance and Stanley Windwer for their advice and helpful comments. We extend our appreciation to the entire staff at Saunders College Publishing, especially to John Vondeling, publisher, and Beth Rosato, development editor, for their encouragement and excellent efforts in producing this second edition.

CONTENTS

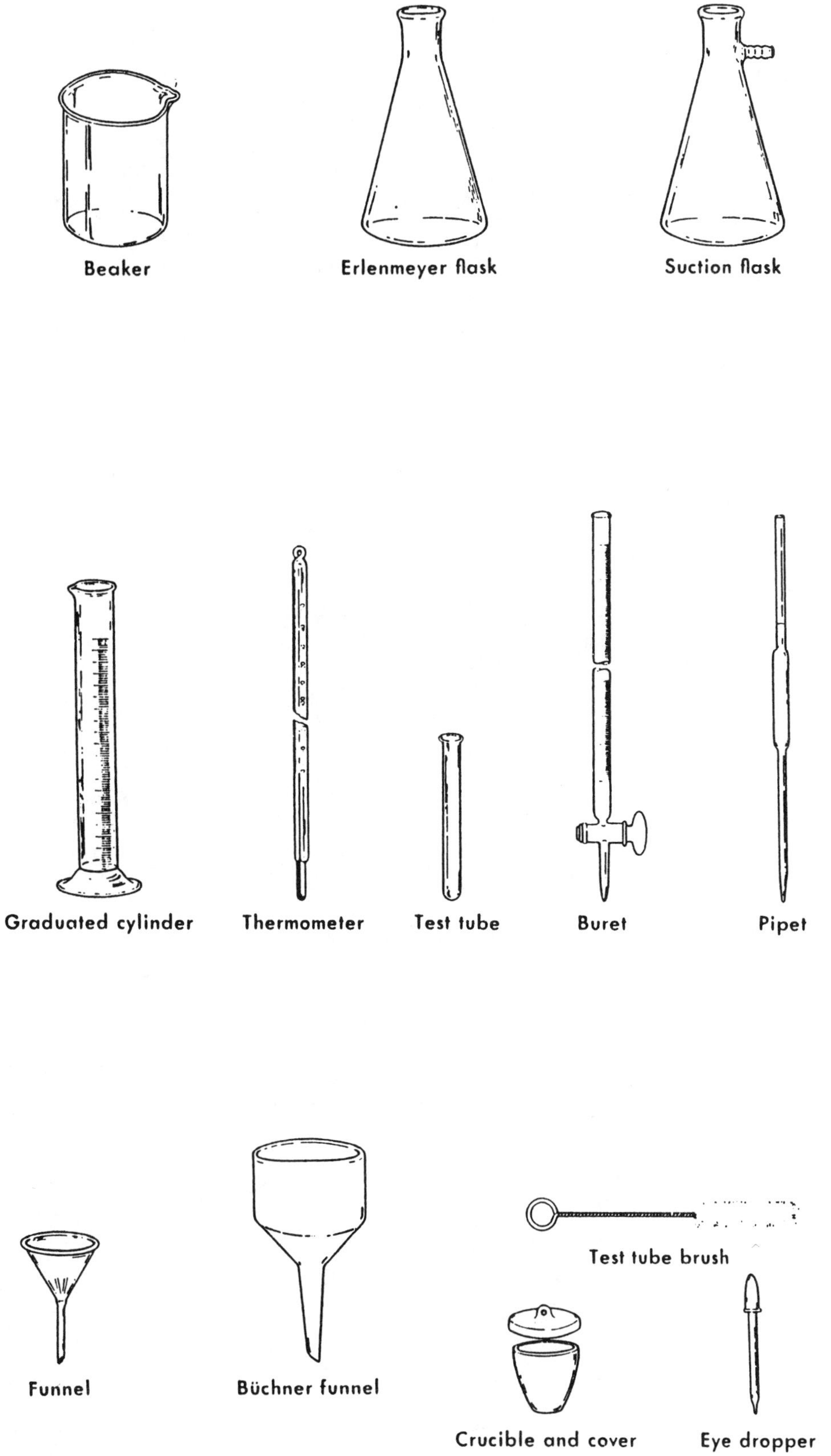

Figure 1 Common laboratory equipment. (From Weiner, S.A., and Peters, E.I.: Introduction to Chemical Principles. W.B. Saunders, Philadelphia, 1980.)

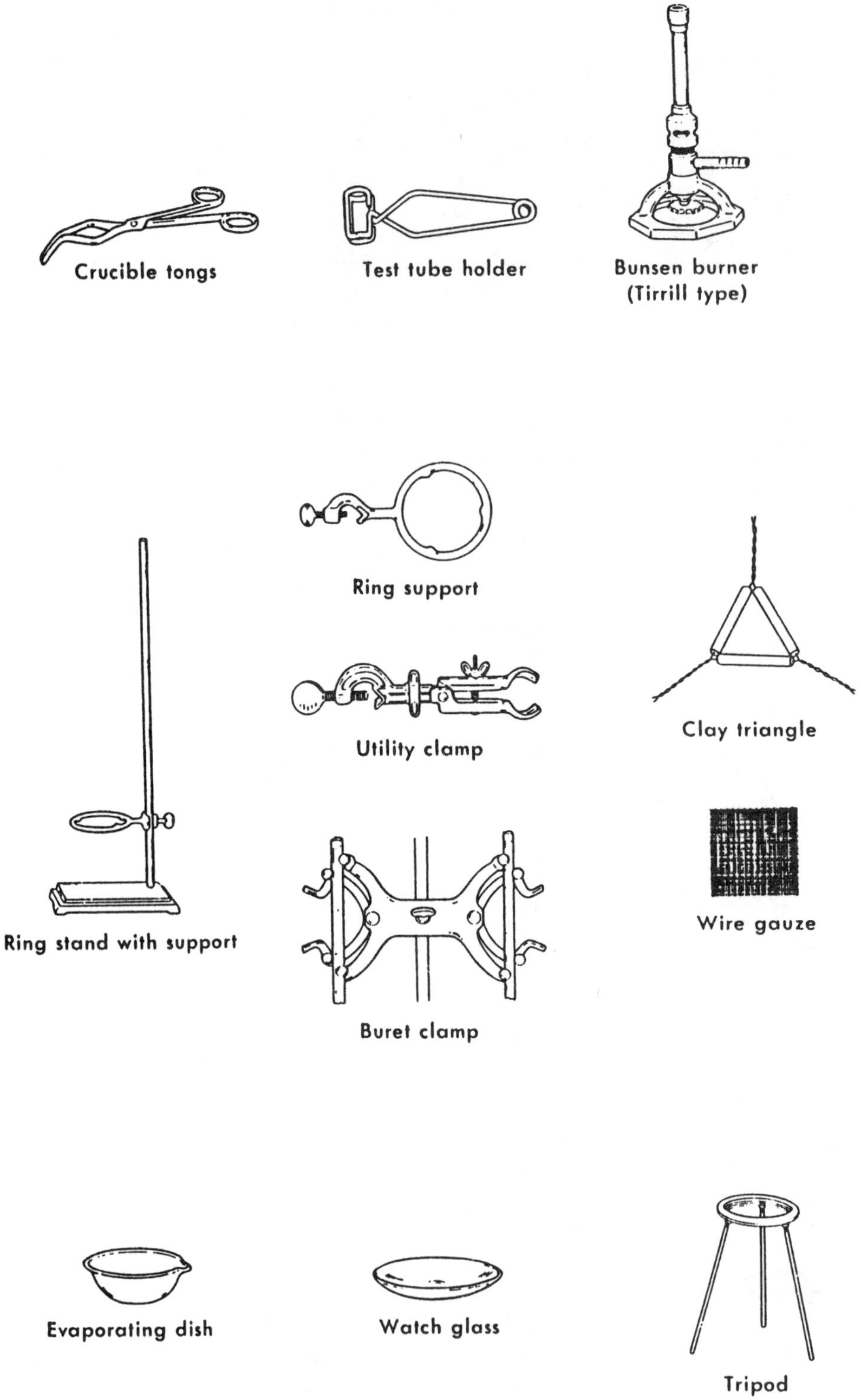
Crucible tongs
Test tube holder
Bunsen burner
(Tirrill type)
Ring support
Ring stand with support
Utility clamp
Clay triangle
Wire gauze
Buret clamp
Evaporating dish
Watch glass
Tripod

Figure 1 (continued)

EXPERIMENT 1

Structure in organic compounds: use of molecular models

Background

The study of organic chemistry usually involves those molecules which contain carbon. Thus, a convenient definition of organic chemistry is the chemistry of carbon compounds.

There are several characteristics of organic compounds that make their study interesting:

a. Carbon forms strong bonds to itself as well as to other elements; the most common elements found in organic compounds, besides carbon, are hydrogen, oxygen and nitrogen.

b. Carbon atoms are generally tetravalent. This means that carbon atoms in most organic compounds are bound by four covalent bonds to adjacent atoms.

c. Organic molecules are three dimensional and occupy space. The covalent bonds which carbon makes to adjacent atoms are at discrete angles to each other. Depending on the type of organic compound, the angle may be 180°, 120°, or 109.5°. These angles correspond to compounds which have triple bonds (1), double bonds (2) and single bonds (3), respectively.

(1)

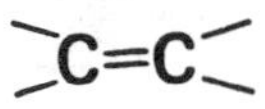

(2)

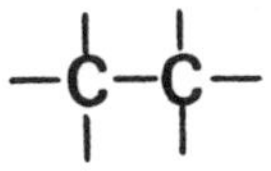

(3)

d. Organic compounds can have a limitless variety in composition, shape and structure.

Thus, while a molecular formula tells the number and type of atoms present in a compound, it tells nothing about the structure. The structural formula is a two-dimensional representation of a molecule and shows the sequence in which the atoms are connected and the bond type. For example, the molecular formula C_4H_{10} can be represented by two different structures: n-butane (4) and isobutane (5) (2-methylpropane).

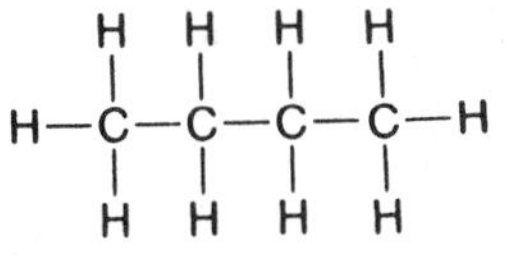

n-Butane (4)

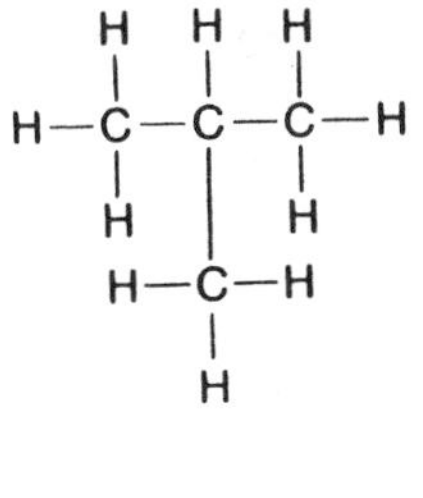

Isobutane (5)
(2-methylpropane)

Consider also the molecular formula C_2H_6O. There are two structures which correspond to this formula: dimethyl ether (6) and ethyl alcohol (7).

$$H-\overset{\displaystyle H}{\underset{\displaystyle H}{\overset{|}{\underset{|}{C}}}}-O-\overset{\displaystyle H}{\underset{\displaystyle H}{\overset{|}{\underset{|}{C}}}}-H \qquad H-\overset{\displaystyle H}{\underset{\displaystyle H}{\overset{|}{\underset{|}{C}}}}-\overset{\displaystyle H}{\underset{\displaystyle H}{\overset{|}{\underset{|}{C}}}}-O-H$$

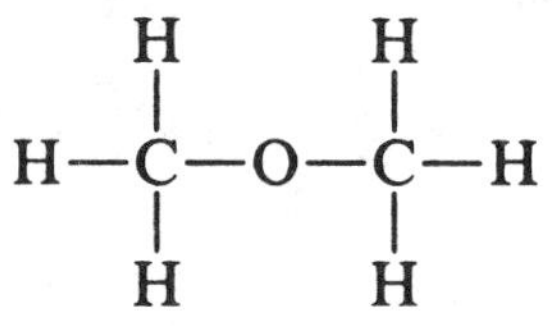

Dimethyl ether (6) Ethyl alcohol (7)

In the pairs above, each structural formula represents a different compound. Each compound has its own unique set of physical and chemical properties. Compounds with the same molecular formula but with different structural formulas are called isomers.

The three-dimensional character of molecular structure is shown through molecular model building. With molecular models the number and types of bonds between atoms and the spatial arrangements of the atoms can be visualized for the molecules. This allows the comparison of isomers for a given set of compounds.

Objectives

1. To use models to visualize structure in organic molecules.
2. To build and compare isomers having a given molecular formula.
3. To explore the three-dimensional character of organic molecules.

Procedure

Obtain from the laboratory instructor a set of ball and stick molecular models. The set contains the following parts:

 a. 2 Black spheres representing Carbon; this tetracovalent element has four holes;
 b. 6 Yellow spheres representing Hydrogen; this monovalent element has one hole;
 c. 2 Colored spheres representing the halogen Chlorine; this monovalent element has one hole;
 d. 1 Blue sphere representing Oxygen; this divalent element has two holes;
 e. 8 Sticks to represent bonds.

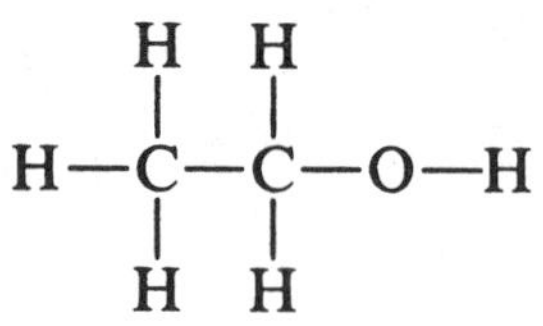

1. With your models construct the molecule methane. Methane is a simple hydrocarbon consisting of one carbon and four hydrogens. After you put the model together, answer the questions below in the appropriate space on the Report Sheet.
   a. With the model resting so that three hydrogens are on the desk, examine the structure. Move the structure so that a different set of three hydrogens are on the desk each time. Is there any difference between the way that the two structures look (1a)?
   b. Does the term "equivalent" adequately describe the four hydrogens of methane (1b)?
   c. Tilt the model so that only two hydrogens are in contact with the desk and imagine pressing the model flat onto the desk top. Draw the way in which the methane molecule would look in two dimensional space (1c). This is the usual way that three dimensional structures are written.
   d. Using a protractor, measure the angle H-C-H on the model (1d).

2. Replace one of the hydrogens of the methane model with a colored sphere which represents the halogen chlorine. The new model is methyl chloride, CH_3Cl. Position the model so that the three hydrogens are on the desk.
   a. Grasp the atom representing chlorine and tilt it to the right, keeping two hydrogens on the desk. Write the structure of the projection on the Report Sheet (2a).
   b. Return the model to its original position and then tilt, as before, but this time to the left. Write this projection on the Report Sheet (2b).
   c. While the projection of the molecule changes, does the structure of methyl chloride change (2c)?

3. Now replace a second hydrogen with another chlorine sphere. The new molecule is dichloromethane, CH_2Cl_2.
   a. Examine the model as you twist and turn it in space. Are the projections given below isomers of the molecule CH_2Cl_2 or representations of the same structure only seen differently in three dimensions (3a)?

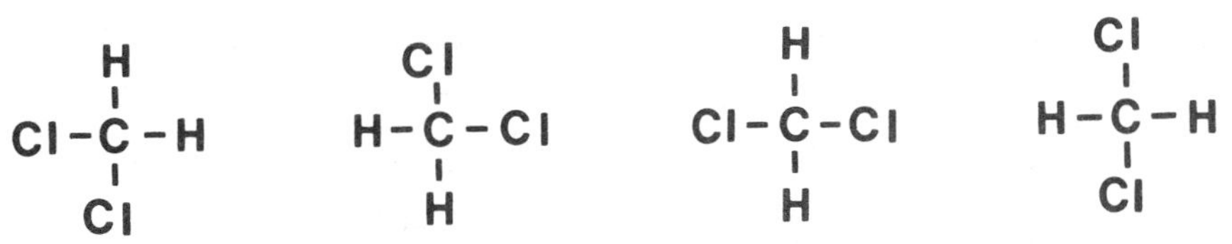

4. Construct the molecule ethane, C_2H_6. Note that you can make ethane from the methane model by removing a hydrogen and replacing the hydrogen with a methyl group, $-CH_3$.
   a. Write the structural formula for ethane (4a).
   b. Are all the hydrogens attached to the carbon atoms equivalent (4b)?
   c. Replace any hydrogen in your model with chlorine. Write the structure of molecule ethyl chloride, C_2H_5Cl (4c).
   d. Twist and turn your model. Write a projection of the ethyl chloride that would be different from 4c (4d).
   e. Do the projections that are written represent different isomers or the same compound (4e)?

5. Dichloroethane, $C_2H_4Cl_2$
   a. In your molecule of ethyl chloride, if you remove another hydrogen, note that you now have a choice among the hydrogens. You can either remove a hydrogen from the carbon to which the chlorine is attached, or you can remove a hydrogen from carbon that has only hydrogens attached.
   b. First remove the hydrogen from the carbon that has the chlorine attached and replace it with a second chlorine. Write its structure on the Report Sheet (5a).
   c. Compare this structure to the model which would result from removal of a hydrogen from the other carbon and its replacement by chlorine. Write its structure (5b) and compare it to the previous example. One isomer is 1,1-dichloroethane; the other is 1,2-dichloroethane. Label the structures drawn on the Report Sheet with the correct name.
   d. Are all the hydrogens of ethyl chloride equivalent (5c)? Are some of the hydrogens equivalent (5d)? Which ones are equivalent to each other (5e)?

6. Butane
   a. Butane has the formula C_4H_{10}. With help from a partner, construct a model of n-butane by connecting the four carbons in a series (C-C-C-C) and then adding the needed hydrogens. First orient the model in such a way that the carbons appear as a straight line. Next tilt the model so that the carbons appear as a zig-zag line. Then twist around any of the C-C bonds so that a part of the chain is at an angle to the remainder. Draw each of these structures in the space on the Report Sheet (6a). Note that the structures you draw are for the same molecule but represent only a different orientation and projection.

b. Examine the structure of n-butane for equivalent hydrogens. In the space on the Report Sheet (6b) redraw the structure of n-butane and label those hydrogens which are equivalent to each other. On the basis of this examination, predict how many monochlorobutane isomers (C_4H_9Cl) that could be obtained (6c). Test your prediction by replacement of hydrogen by chlorine on the models. Draw the structures of these isomers (6d).

c. Reconstruct the butane system. First form a three carbon chain, then connect the fourth carbon to the center carbon of the three carbon chain. Add the necessary hydrogens. Draw the structure of isobutane (6e). Can any manipulation of the model, by twisting or turning of the model or by rotation of any of the bonds, give you the n-butane system? If these two, n-butane and isobutane, are isomers, then how may we recognize that any two structures are isomers (6f)?

d. Examine the structure of isobutane for equivalent hydrogens. In the space on the Report Sheet (6g), redraw the structure of isobutane and identify the equivalent hydrogens. Predict how many monochloroisobutanes could be formed (6h) and test your prediction by replacement of hydrogen by chlorine on the model. Draw the structures of these isomers (6i).

7. C_2H_6O

a. There are two isomers with the molecular formula C_2H_6O, ethyl alcohol and dimethyl ether. With your partner, construct both of these isomers. Draw these isomers on the Report Sheet (7a) and name each one.

b. Manipulate each model. Can either be turned into the another by a simple twist or turn (7b)?

c. For each compound, label those hydrogens which are equivalent. How many sets of equivalent hydrogens are there in ethyl alcohol and dimethyl ether (7c)?

8. Optional: Butene

a. If springs are available for the construction of double bonds, construct 2-butene, CH_3-CH=CH-CH_3.

b. There are two isomers for compounds of this formulation: the isomer with the 2 CH_3 groups on the same side of the double bond, cis-2-butene; the isomer with the 2 CH_3 groups on opposite sides of the double bond, trans-2-butene. Draw these two structures on the Report Sheet (8a).

c. Can you twist, turn or rotate one model into the other? Explain (8b).

d. How many bonds are connected to any single carbon of these structures (8c)?

e. With the protractor measure the C-C=C angle (8d).

Chemicals and Equipment

1. Molecular models consisting of
   2 Black spheres
   6 Yellow spheres
   2 Colored spheres (e.g. green)
   1 Blue sphere
   8 Sticks
2. Protractor
3. Optional: 2 springs.

NAME SECTION DATE

PARTNER GRADE

Experiment 1

PRE-LAB QUESTIONS

1. How many bonds can each of the elements below form with neighboring atoms in a compound?

   C H O N Br

2. What is a convenient definition for organic chemistry?

3. Write structural formulas for the two compounds with molecular formula C_3H_7Cl.

4. How are the compounds in no. 3 (above) related?

NAME SECTION DATE

PARTNER GRADE

Experiment 1

REPORT SHEET

1. Methane
   a.

   b.

   c.

   d.

2. Methyl chloride
   a.

   b.

   c.

3. Dichloromethane
   a.

4. Ethane and ethyl chloride
    a.
    b.
    c.
    d.
    e.

5. Dichloroethane
    a.
    b.
    c.
    d.
    e.

6. Butane
   a.

   b.

   c.

   d.

   e.

   f.

   g.

   h.

   i.

7. C_2H_6O
   a.

   b.

   c. Ethyl alcohol has __________ set(s) of equivalent hydrogens.

   Dimethyl ether has __________ set(s) of equivalent hydrogens.

8. Butene
   a.

   b.

   c.

   d. C-C=C angle

POST-LAB QUESTIONS

1. Propane has the molecular formula C_3H_8. Write a structural formula for this compound.

2. Redraw the structure of propane and identify equivalent hydrogens. Identify equivalent sets by letters, e.g., H_a, H_b, etc.

3. Draw the structural formulas for the four (4) isomers of C_4H_9Br.

4. Draw the structural formulas for the four (4) isomers of the butenes, C_4H_8.

EXPERIMENT 2
Identification of hydrocarbons

Background

The number of known organic compounds total into the millions. Of these compounds the simplest types are those which contain only hydrogen and carbon atoms. These are known as <u>hydrocarbons</u>. Because of the number and variety of hydrocarbons that can exist, some means of classification is necessary.

One means of classification depends on the way in which carbon atoms are connected. <u>Aliphatic</u> hydrocarbons are compounds consisting of carbons linked either in a single chain or in a branched chain. <u>Cyclic</u> hydrocarbons are compounds having the carbon atoms linked in a closed polygon. For example, <u>n</u>-hexane and 2-methylpentane are aliphatic molecules, while cyclohexane is a cyclic system.

$CH_3CH_2CH_2CH_2CH_2CH_3$ $\qquad$ $CH_3CHCH_2CH_2CH_3$
$\qquad\qquad\qquad\qquad\qquad\qquad\qquad\qquad\quad |$
$\qquad\qquad\qquad\qquad\qquad\qquad\qquad\qquad\ CH_3$

$\qquad$ <u>n</u>-Hexane $\qquad\qquad\qquad$ 2-Methylpentane $\qquad\qquad\qquad$ Cyclohexane

Another means of classification depends on the type of bonding that exists between carbons. Hydrocarbons which contain only carbon to carbon single bonds are called <u>alkanes</u>. These are also referred to as <u>saturated</u> molecules. Hydrocarbons containing at least one carbon to carbon double bond are called <u>alkenes</u>, and those compounds with at least one carbon to carbon triple bond are <u>alkynes</u>. These are compounds that are referred to as <u>unsaturated</u> molecules. Finally, a class of cyclic hydrocarbons which have a high degree of unsaturation are called <u>aromatic</u>. Table 2.1 distinguishes between the families of hydrocarbons.

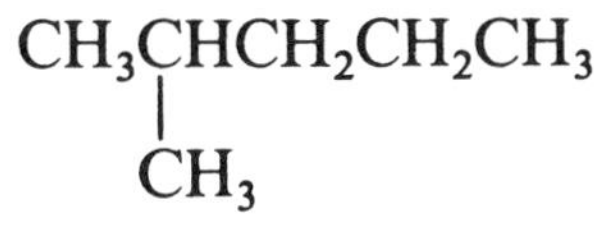

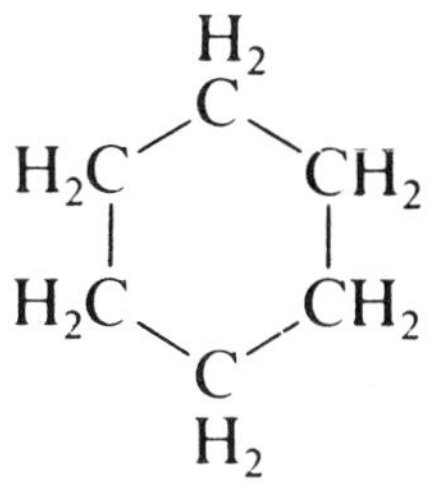

Table 2.1 Types of Hydrocarbons

Class	Characteristic Bond Type		Example	
I. Aliphatic				
1. Alkane[a]	$-\overset{\vert}{\underset{\vert}{C}}-\overset{\vert}{\underset{\vert}{C}}-$	single	$CH_3CH_2CH_2CH_2CH_2CH_3$	n-hexane
2. Alkene[b]	$>C=C<$	double	$CH_3CH_2CH_2CH_2CH{=}CH_2$	1-hexene
3. Alkyne[b]	$-C\equiv C-$	triple	$CH_3CH_2CH_2CH_2C\equiv CH$	1-hexyne
II. Cyclic				
1. Cycloalkane[a]	$-\overset{\vert}{\underset{\vert}{C}}-\overset{\vert}{\underset{\vert}{C}}-$	single	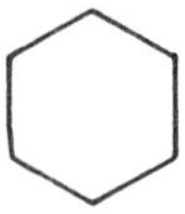	cyclohexane
2. Cycloalkene[b]	$>C=C<$	double		cyclohexene
3. Aromatic	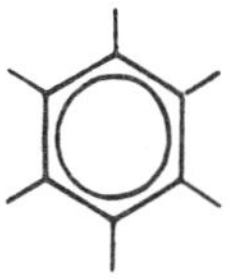			benzene
			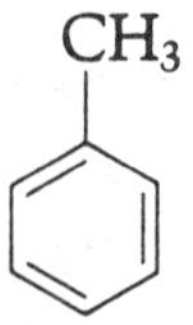	toluene

[a]Saturated [b]Unsaturated

With so many compounds possible, identification of the bond type is an important step in establishing the molecular structure. Quick, simple tests on small samples can establish the physical and chemical properties of the compounds by class.

Some of the observed physical properties of hydrocarbons result from the non-polar character of the compounds. In general, hydrocarbons do not mix with polar solvents such as water or ethyl alcohol. On the other hand, hydrocarbons mix with relatively non-polar solvents such as ligroine (a mixture of alkanes), carbon tetrachloride or dichloromethane. Since the density of most carbon-hydrogen containing compounds is less than that of water, they will float. Crude oil and crude oil products (home heating oil and gasoline) are mixtures of hydrocarbons; these substances, when spilled on water, spread quickly along the surface because they are insoluble in water.

The chemical reactivity of hydrocarbons is determined by the type of bond in the compound. While saturated hydrocarbons, alkanes, will burn (undergo combustion), they are generally unreactive to most reagents. Unsaturated hydrocarbons, alkenes and alkynes, not only burn, but also react by addition of reagents to the double or triple bonds. The addition products become saturated, with fragments of the reagent becoming attached to the carbons of the multiple bond. Aromatic compounds, with a higher carbon to hydrogen ratio than non-aromatic compounds, burn with a sooty flame as a result of unburned carbon particles being present. The reaction of these compounds with reagents is by substitution rather than by addition.

1. Combustion. The major component in "natural gas" is the hydrocarbon methane. Other hydrocarbons used for heating or cooking purposes are propane and butane. The products from combustion are carbon dioxide and water (heat is evolved, also).

$$CH_4 + 2O_2 \longrightarrow CO_2 + 2H_2O$$

$$(CH_3)_2CHCH_2CH_3 + 8O_2 \longrightarrow 5CO_2 + 6H_2O$$

2. Reaction with bromine. Unsaturated hydrocarbons react rapidly with bromine in a solution of carbon tetrachloride or hexane. The reaction is the addition of the elements of bromine to the carbons of the multiple bonds.

$$CH_3CH{=}CHCH_3 + Br_2 \longrightarrow \underset{\text{Colorless}}{CH_3\overset{\overset{\displaystyle Br}{|}}{C}H-\overset{\overset{\displaystyle Br}{|}}{C}HCH_3}$$

Red Colorless

$$CH_3C{\equiv}CCH_3 + 2\,Br_2 \longrightarrow CH_3\overset{\overset{\displaystyle Br}{|}}{\underset{\underset{\displaystyle Br}{|}}{C}}-\overset{\overset{\displaystyle Br}{|}}{\underset{\underset{\displaystyle Br}{|}}{C}}CH_3$$

Red Colorless

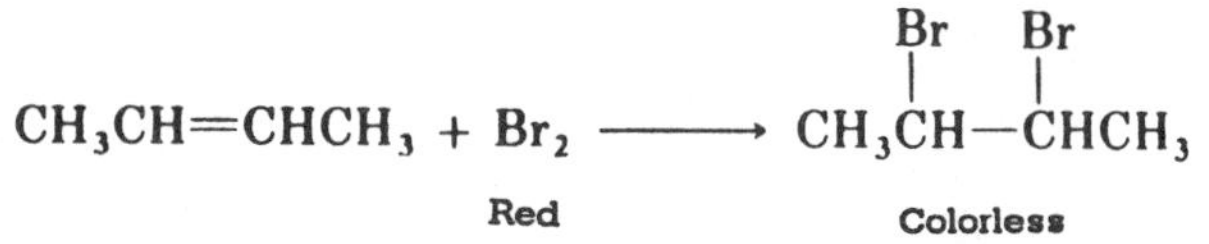

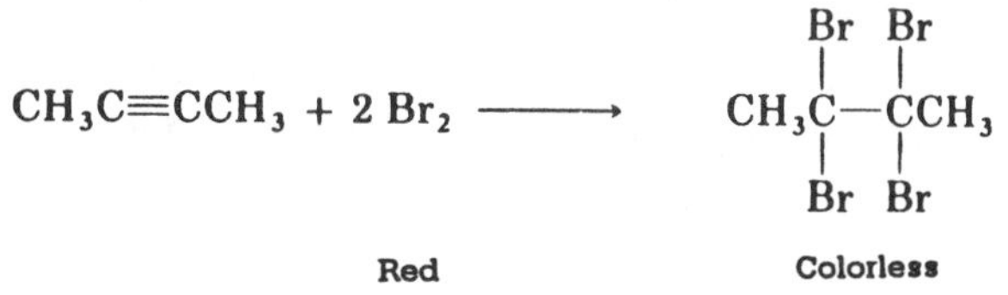

The bromine solution is brown; the product which has the bromine atoms attached to carbon is colorless. Thus, a reaction has taken place when there is a loss of color from the bromine solution and a colorless solution remains. Since alkanes have only single C-C bonds present, no reaction with bromine is observed; the brown color of the reagent would persist when added. Aromatic compounds resist addition reactions because of their unique characteristics. These compounds react with bromine in the presence of a catalyst such as iron filings or aluminum chloride.

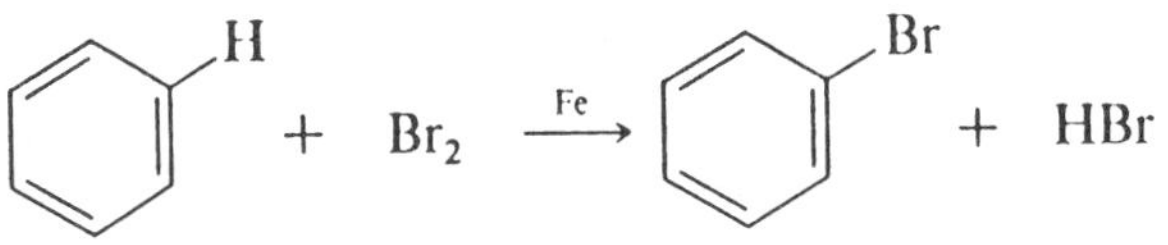

Note that a substitution reaction has taken place and the gas HBr is produced.

3. Reaction with concentrated sulfuric acid. Unsaturated hydrocarbons react with cold concentrated sulfuric acid by addition. Alkyl sulfonic acids form as products and are soluble in H_2SO_4.

$$CH_3{-}CH{=}CH{-}CH_3 + \underset{(H_2SO_4)}{HOSO_2OH} \rightarrow CH_3{-}\underset{\displaystyle H}{\underset{|}{CH}}{-}\underset{\displaystyle OSO_2OH}{\underset{|}{CH}}{-}CH_3$$

Saturated hydrocarbons are unreactive, while alkynes and aromatic compounds react very slowly.

4. Reaction with potassium permanganate. Dilute or alkaline solutions of $KMnO_4$ oxidize unsaturated compounds. Alkanes and aromatic compounds are generally unreactive. Evidence that a reaction has occurred is by the loss of the purple color of $KMnO_4$ and the formation of the brown precipitate manganese dioxide, MnO_2.

$$3CH_3{-}CH{=}CH{-}CH_3 + \underset{\text{Purple}}{2KMnO_4} + 4H_2O \rightarrow 3CH_3{-}\underset{\displaystyle OH}{\underset{|}{CH}}{-}\underset{\displaystyle OH}{\underset{|}{CH}}{-}CH_3 + \underset{\text{Brown}}{2MnO_2} + 2KOH$$

Note that the product formed from an alkene is a glycol or 1,2-diol.

Objectives

1. To investigate the physical properties, solubility and density, of some hydrocarbons.
2. To compare the chemical reactivity of an alkane, an alkene and an aromatic compound.
3. To use physical and chemical properties to identify an unknown.

Procedure

CAUTION! Assume the organic compounds are highly flammable. Use only small quantities. Keep away from open flames. Assume the organic compounds are toxic and can be absorbed through the skin. Avoid contact; wash if any chemical spills on your person. Handle concentrated sulfuric acid carefully. Flush with water if any spills on your person. Potassium permanganate and bromine are toxic. Although the solutions are dilute, they may cause burns to the skin.

General instructions

1. The hydrocarbons hexane, cyclohexene and toluene (alkane, alkene and aromatic) are available in dropper bottles.

2. The reagents 1% Br_2 in hexane, 1% aqueous $KMnO_4$ and concentrated H_2SO_4 are available in dropper bottles.

3. Unknowns are in dropper bottles labeled A, B and C. They may include an alkane, an alkene or an aromatic compound.

4. Record all data and observations in the appropriate places on the Report Sheet.

5. Dispose of all organic wastes as directed by the instructor. Do not pour into the sink!

Physical properties of hydrocarbons

1. A test tube of 100 x 13 mm will be suitable for this test. When mixing the components, grip the test tube between thumb and forefinger; it should be held firmly enough to keep from slipping but loosely enough so that when the third and fourth fingers tap it, the contents will be agitated enough to mix.

2. Water solubility of hydrocarbons. Label six test tubes with the name of the substance to be tested. Place into each test tube 10 drops of the appropriate hydrocarbon: hexane, cyclohexene, toluene, unknown A, unknown B, unknown C. Add about 10 drops of water dropwise into each test tube. Is there any separation of components? Which component is on the bottom; which component is on the top? Mix the contents as described above. What happens when the contents are allowed to settle? Record your observations. Save these solutions for comparison with the next part.

3. Solubility of hydrocarbons in ligroine. Label six test tubes with the name of the substance to be tested. Place into each test tube 10 drops of the appropriate hydrocarbon: hexane, cyclohexene, toluene, unknown A, unknown B, unknown C. Add about 10 drops of ligroine dropwise into each test tube. Is there a separation of components? Is there a bottom layer and top layer? Mix the contents as described above. Is there any change in the appearance of the contents before and after mixing? Compare these test tubes to those from the previous part. Record your observations.

Chemical properties of hydrocarbons

1. Combustion. The instructor will demonstrate this test in the fume hood. Place 10 drops of each hydrocarbon and unknown on separate watch glasses. Carefully ignite each sample with a match. Observe the flame and color of the smoke for each of the samples. Record your observations on the Report Sheet.

2. Reaction with bromine. Label six clean, dry test tubes with the name of the substance to be tested. Place into each test tube 20 drops of the appropriate hydrocarbon: hexane, cyclohexene, toluene, unknown A, unknown B, unknown C. Carefully add, dropwise and with shaking, 1% Br_2 in hexane. (**CAUTION:** Use in hood.) Keep count of the number of drops needed to have the color persist; do not add more than 20 drops. Record your observations.

To the test tube containing toluene, add 10 more drops of 1% Br_2 solution and a small quantity of iron filings; shake the mixture. Place a piece of moistened blue litmus paper on the test tube opening. Record any change in the color of the solution and the litmus paper.

3. Reaction with $KMnO_4$. Label six clean, dry test tubes with the name of the substance to be tested. Place into each test tube 20 drops of the appropriate hydrocarbon: hexane, cyclohexene, toluene, unknown A, unknown B, unknown C. Carefully add, dropwise and with shaking 1% aqueous $KMnO_4$ solution. Keep count of the number of drops needed to have the color persist; do not add more than 20 drops. Record your observations.

4. Reaction with concentrated H_2SO_4. Label six clean, dry test tubes with the name of the substance to be tested. Place into each test tube 20 drops of the appropriate hydrocarbon: hexane, cyclohexene, toluene, unknown A, unknown B, unknown C. Place all of the test tubes in an ice bath. Carefully add with shaking, 10 drops of cold, concentrated sulfuric acid to each test tube. Note whether heat is evolved by feeling the test tube. Note whether the solution has become homogeneous or whether a color is produced. (The evolution of heat or the formation of a homogeneous solution or the appearance of a color is evidence that a reaction has occurred.) Record your observations.

5. Unknowns. By comparing the observations you make for your unknowns with that of the known hydrocarbons, you can identify unknowns A, B and C. Record their identities on your Report Sheet.

Chemicals and Equipment

1. Hexane
2. Cyclohexene
3. Toluene
4. Ligroine
5. 1% Br_2 in hexane
6. 1% aqueous $KMnO_4$
7. Concentrated H_2SO_4
8. Iron filings or powder
9. Blue litmus paper
10. Watch glasses
11. Unknowns A, B, and C
12. Ice

NAME SECTION DATE

PARTNER GRADE

Experiment 2

PRE-LAB QUESTIONS

1. Why is n-hexane known as an hydrocarbon?

2. Show the structural feature that distinguishes whether an hydrocarbon is an

   alkane

   alkene

   alkyne

   aromatic

3. Hydrocarbons do not mix with water. What accounts for this property?

NAME SECTION DATE

PARTNER GRADE

Experiment 2

REPORT SHEET

Physical properties of hydrocarbons

Hydrocarbon	Solubility		Density
	H_2O	Ligroine	
Hexane			
Cyclohexene			
Toluene			
Unknown A			
Unknown B			
Unknown C			

Chemical properties of hydrocarbons

Hydrocarbon	Combustion	Bromine Test	$KMnO_4$ Test	H_2SO_4
Hexane				
Cyclohexene				
Toluene				
Unknown A				
Unknown B				
Unknown C				

Unknown A is______________________________.

Unknown B is______________________________.

Unknown C is______________________________.

POST-LAB QUESTIONS

1. Write the structure of the major organic product for the following reactions:

   a. $CH_3{-}CH{=}CH_2 + Br_2 \rightarrow$

b. 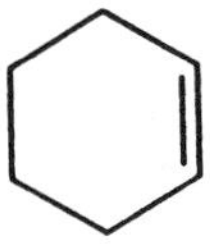 $+ Br_2 \rightarrow$

c. $CH_3{-}CH{=}CH{-}CH_3 + KMnO_4 + H_2O \rightarrow$

d. $CH_3CH_2{-}CH{=}CH{-}CH_2CH_3 + H_2SO_4 \rightarrow$

2. Of the compounds used in this experiment (hexane, cyclohexene, toluene) which was the most reactive with bromine solution?

3. On the basis of the chemical tests used in this experiment, is it possible to <u>clearly</u> distinguish between an alkane and an alkene?

EXPERIMENT 3

Column and paper chromatography: Separation of plant pigments.

Background

Chromatography is a widely used experimental technique by which a mixture of compounds can be separated into its individual components. Two kinds of chromatographic experiments will be explored. In column chromatography a mixture of components dissolved in a solvent is poured over a column of solid adsorbent and is eluted with the same or a different solvent. This is therefore a solid-liquid system, the stationary phase (the adsorbent) is solid and the mobile phase (the eluent) is liquid. In paper chromatography the paper adsorbs water from the atmosphere of the developing chromatogram. (The water is present in the air as vapor and it may be supplied as one component in the eluting solution). The water is the stationary phase. The (other) components of the eluting solvent is the mobile phase and carries with it the components of the mixture. This is a liquid-liquid system.

Column chromatography is used most conveniently for preparative purposes, when one deals with a relatively large amount of the mixture and the components need to be isolated in mg or g quantities. Paper chromatography on the other hand, is used mostly for analytical purposes. Microgram or even picogram quantities can be separated by this technique and they can be characterized by their R_f number. This number is an index of how far a certain spot moved on the paper.

$$R_f = \frac{\text{Distance of the center of the sample spot from the origin}}{\text{Distance of the solvent front from the origin}}$$

For example in Figure 23.1 the R_f values are as follows:

R_f (substance 1)= 3.1 cm/11.2 cm= 0.28 and

R_f (substance 2)= 8.5 cm/11.2 cm= 0.76

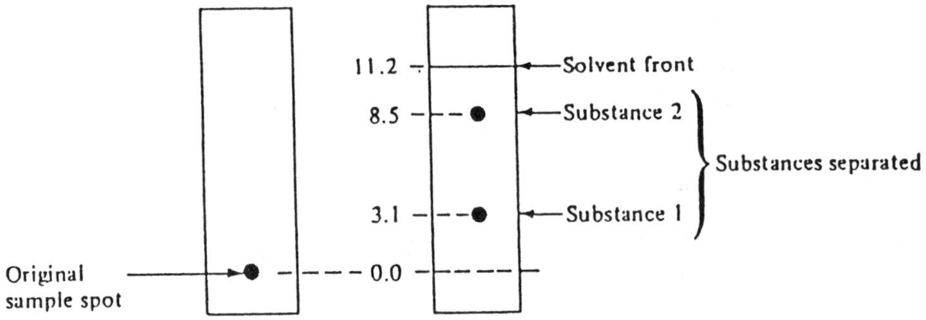

Figure 3.1 Illustration of chromatograms before and after elution

Using the R_f values one is able to identify the components of the mixture with the individual components. The two main pigment components of tomato paste are β-carotene (yellow-orange) and lycopene (red) pigments. Their structures are given below:

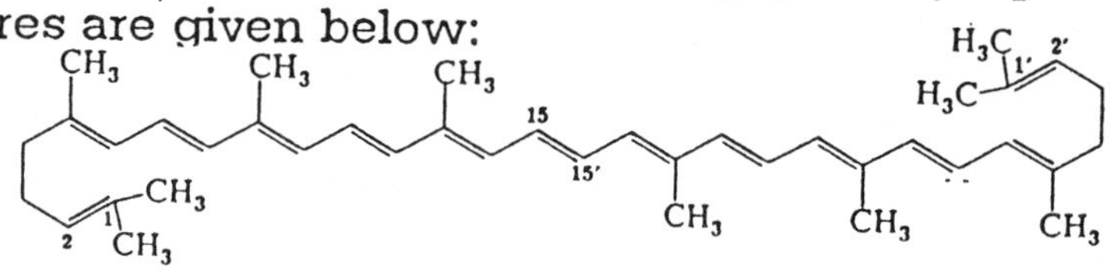

Lycopene

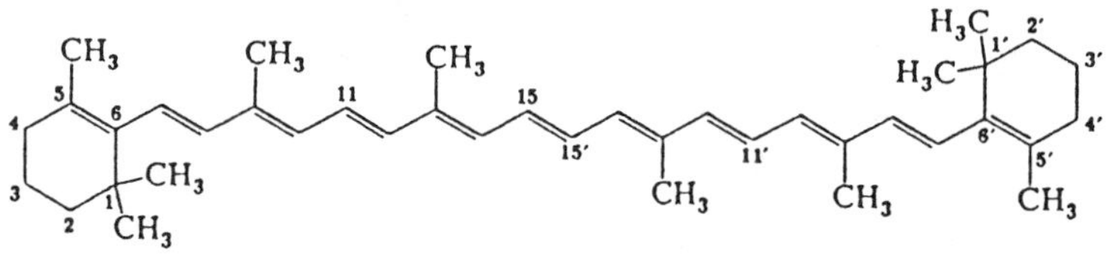

β-Carotene

The colors of these pigments are due to the numerous double bonds in their structure. When bromine is added to double bonds it saturates them and the color changes accordingly. In the tomato juice "rainbow" experiment we stir bromine water into the tomato juice. The slow stirring allows the bromine water to penetrate deeper and deeper into the cylinder in which the tomato juice was placed. As the bromine penetrates more and more double bonds will be saturated. Therefore, you may be able to observe a continuous change, a "rainbow" of colors starting with the reddish tomato color at the bottom of the cylinder where no reaction occurred since the bromine did not reach the bottom. Lighter colors will be observed on the top of the cylinder where most of the double bonds have been saturated.

Objective

1. To compare separation of components of a mixture by two different techniques.
2. To demonstrate the effect of bromination on plant pigments of tomato juice.

Procedure

Part A. Paper chromatography

1.Obtain a sheet of Whatman no.1 filter paper, cut to size.

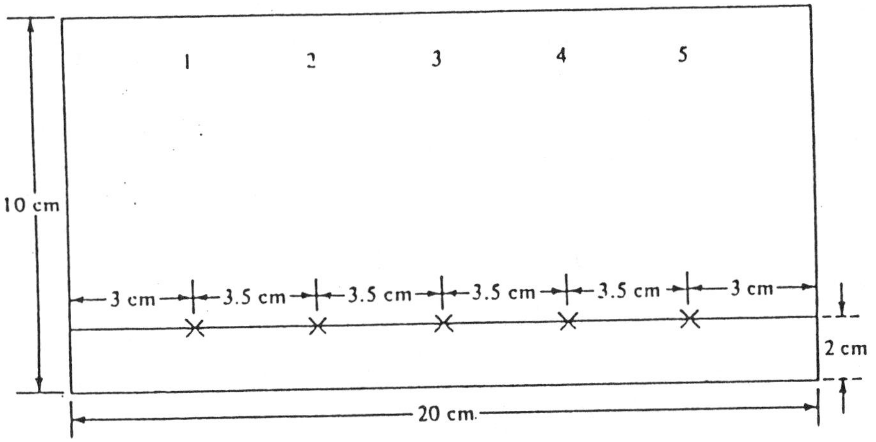

Figure 3.2 Preparation of chromatographic paper for spotting.

2.Plan the **spotting** of the samples as illustrated on Fig. 3.2. Five spots will be applied. The first and fifth will be β-carotene solutions supplied by your instructor. The second, third and fourth lanes will have your tomato paste extracts in different concentrations. Use a pencil to mark lightly the spots according to Fig. 3.2.

3.Pigments of tomato paste will be **extracted** in two steps.

(a)Weigh about 10 g of tomato paste in a 50 mL beaker. Add 15 mL of 95 % ethanol. Stir the mixture vigorously with a spatula until the paste will not stick to the stirrer. Place a small amount of glass wool (the size of a pea) in a small funnel blocking the funnel exit. Place the funnel into a 50 mL Erlenmayer flask and pour the tomato paste - ethanol mixture into the funnel. When the filtration is completed squeeze the glass wool lightly with your spatula. In this step we removed the water from the tomato paste and the aqueous components are in the filtrate, which we discard. The residue in the glass wool will be used to extract the pigments.

(b) Place the residue in the glass wool in a 50 mL beaker. Add 10 mL dichloromethane and stir the mixture for about two minutes to extract the pigments. Filter the extract as before through a new funnel with glass wool blocking the exit, into a new and clean 50 mL beaker. Place the beaker under the hood on a hot plate (or water bath) and evaporate it to about 1 mL volume. Use low heat and take care not to evaporate all the solvent. After evaporation cover the beaker with aluminum foil.

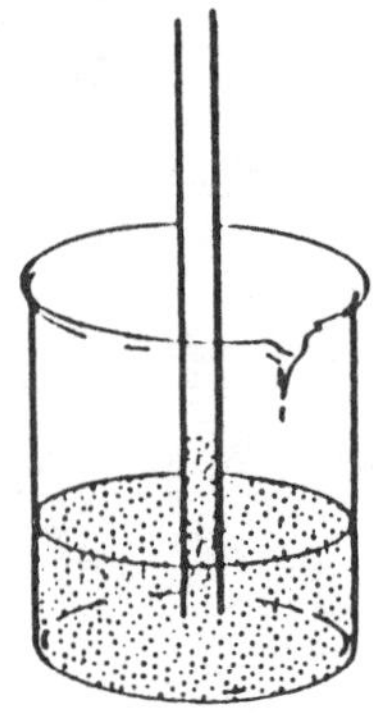

Figure 3.3 Withdrawing samples by a capillary tube.

4. Spotting. Place your chromatographic paper on a clean area (another filter paper) in order not to contaminate it. Use separate capillaries for your tomato paste extract and for the β- carotene solution. First apply your capillary to the extracted pigment by dipping it into the solution as illustrated in Fig. 3.3. Apply the capillary lightly to the chromatographic paper by touching sequentially the spots marked 2, 3 and 4. Make sure you apply only small spots, not larger than 2 mm diameter, by quickly withdrawing the capillary from the paper each time you touch it. (See Fig. 3.4).

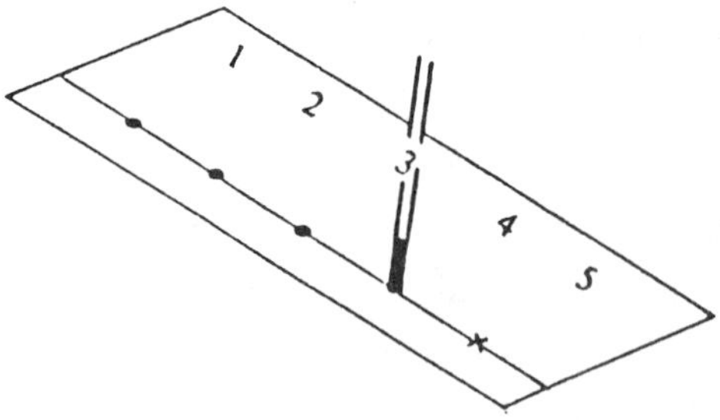

Figure 3.4 Spotting.

While allowing the spots to dry, use your second capillary to apply spots of β- carotene in lanes 1 and 5. Return to the first capillary and apply another spot of the extract on top of the spots of lanes 3 and 4. Let them dry (Fig. 3.5). Finally apply one more spot on top of lane 4. Let the spots dry. The unused extract in your beaker should be covered with aluminum foil. Place it in your drawer in the dark to save it for the second part of your experiment.

Figure 3.5 Drying chromatographic spots.

5. **Developing the Paper Chromatogram.** Curve the paper into a cylinder and staple the edges above the 2 cm line as it is shown in Fig. 23.6.

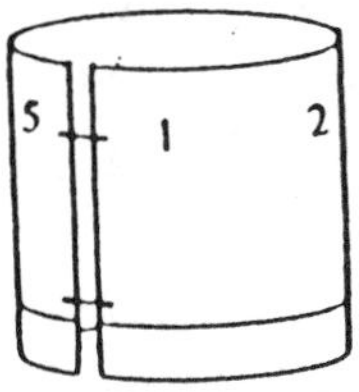

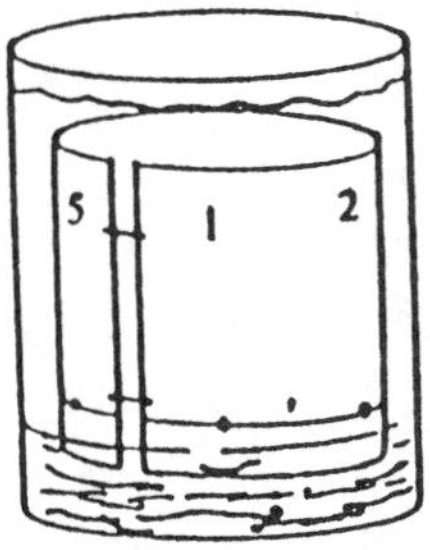

Figure 3.6 Stapling.

Figure 3.7 Developing the chromatogram.

6. Pour 20 mL of the eluting solvent (petroleum ether : toluene : acetone in 45:1:5 ratio, supplied by your instructor) into a 600 mL beaker.

7. Place the stapled chromatogram into the 600 mL beaker the spots being at the bottom near the solvent surface but **not covered by it.** Cover the beaker with aluminum foil. Allow the solvent front to migrate up to 0.5-1 cm below the edge of the paper. This may take from 15 minutes to one hour. Make certain by frequent inspection that the **solvent front does not run over the edge of the paper.** Remove the chromatogram from the beaker when the solvent front reaches 0.5-1 cm from the edge. Proceed to step **11.**

Part B.

8. **Column Chromatography.** While you are waiting for the chromatogram to develop (step 7), you can perform the column chromatography experiment. Take a 25 mL buret. (You may use a chromatograhic column, if available, of 1.6 cm diameter and about 13 cm long, see Fig. 23.8. If you use the column all subsequent quantities below should be doubled).

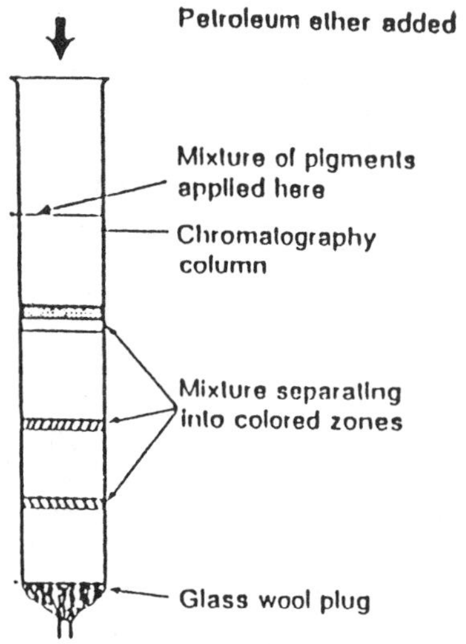

Figure 3.8 Chromatographic column.

Add a small piece of glass wool and with the aid of a glass rod push it down near the stopcock. Add 15-16 mL of petroleum ether to the buret. Open the stopcock slowly and allow the solvent to fill the tip of the buret. Close the stopcock. You should have 12-13 mL of solvent above the glass wool. Weigh 20 g of aluminum oxide (alumina) in a 100 mL beaker. Place a small funnel on top of your buret. Pour the alumina into the buret. Allow the alumina to settle to form a 20 cm column. Drain the solvent but do not allow the column to run dry. Always have at least a 0.5 mL of clear solvent on top of the column. If alumina adheres to the walls of the buret wash it down with more solvent.

9. Transfer by pipet 0.5-1 mL of the extract you stored in your drawer onto the column. The pipet containing the extract should be placed near the surface of the solvent on top of the column. Touching the walls of the buret with the tip of the pipet allow the extract to drain slowly on top of the column. Open the stopcock slightly. Allow the sample to enter the column, but make sure there is a small amount of solvent on top of the column.(The column should never run dry). Add 10 or more mL petroleum ether and wash the sample into the column by opening the stopcock and collecting the eluted solvent in a beaker.

10. As the solvent elutes the sample, you observe the migration of the pigment and their separation into at least two bands. When the fastest moving pigment band reaches near the bottom of the column, close the stopcock and observe the color of the pigment bands and how far they migrated from the top of the column. This concludes the column chromatographic part of the experiment. Discard your solvent in a bottle supplied by your instructor for a later redistillation.

11. Meanwhile your paper chromatogram has developed. You must remove the filter paper from the 600 mL beaker before the solvent front reaches the edges of the paper. Mark the position of the solvent front with a pencil. Put the paper standing on its edges under the hood and let it dry.

Part C.

12. **Tomato Juice "Rainbow".** While waiting for the paper to dry you can perform the following short experiment. Weigh about 15 g tomato paste in a beaker. Add about 30 mL of water and stir. Transfer the tomato juice into a 50 mL cylinder and with the aid of a pipet add 5 mL of saturated bromine water dropwise. With a glass rod stir very gently the solution. Observe the colors and their positions in the cylinder.

Part A cont.

13. Remove the staples from the dried chromatogram. Mark the spots of the pigments by circling it with a pencil. Note the colors of the spots. Measure the distance of the center of each spot from its origin. Calculate the R_f values.

14. If the spots on the chromatogram are faded we can visualize them by exposing the chromatogram to iodine vapor. Place your chromatogram into a wide mouth jar containing a few iodine crystals. Cap the jar and warm it slightly on a hot plate to enhance the sublimation of iodine. The iodine vapor will interact with the faded pigment spots and make them visible. After a few minutes exposure to iodine vapor remove the chromatogram and mark the spots **immediately** with pencil. The spots will fade again with exposure to air. Measure the distance of the center of the spots from the origin and calculate the R_f values.

Chemicals and Equipment

1. Melting point capillaries open at both ends
2. 25 mL buret or chromatographic column
3. Glass wool
4. Whatman no.1 filter papers, 10 x 20 cm, cut to size
5. Heat lamp (optional)
6. Stapler
7. Hot plate with or without water bath
8. Tomato paste
9. Aluminum oxide (alumina)
10. Petroleum ether (b.p. 30-60°C)
11. 95 % ethanol
12. Toluene
13. Acetone
14. 0.5% β-carotene in petroleum ether
15. Saturated bromine water.
16. Iodine crystals
17. Ruler

Name | Section | Date

Partner | Grade

Experiment 3

PRE-LAB QUESTIONS

1. (a) What is the "stationary phase" in column chromatography?
   (b) What is the "mobile phase" in column chromatography?

2. The structures of the two main pigments, lycopene and β-carotene are given in the first part (Background):
   (a) Are these pigments hydrocarbons?

   (b) What functional groups are present in these pigments?

   (c) What solvents will be good for these pigments: polar or non-polar?

3. Write the structure of β-carotene after it completely reacted with Br_2.

Name ______________________ Section ________ Date ________

Partner ______________________ Grade ________

Experiment 3

REPORT SHEET

Paper chromatography

Sample	Distance from origin to solvent front (cm) (a)	Distance from origin to center of spot (cm) (b)	R_f (b)/(a)	Color
β-carotene				
lane 1				
lane 5				
Tomato extract				
lane 2 (a)				
(b)				
(c)				
(d)				
lane 3 (a)				
(b)				
(c)				
(d)				
lane 4 (a)				
(b)				
(c)				
(d)				

Column chromatography

Number of bands	Distance migrated from top of the column (cm)	Color
1		
2		
3		

Describe the colors observed in the tomato juice "rainbow" experiment, starting from the bottom of the cylinder:

1. red 2. 3.

4. 5. 6.

POST-LAB QUESTIONS

1. Did your tomato paste contain lycopene? What support is there for your answer?

2. Which chromatographic technique gave you a better separation? Explain!

3. What is the effect of the amount of sample applied to the paper on the separation of the tomato pigments? Compare the results on lanes 2, 3 and 4 of the paper chromatogram.

4. Based on the R_f value you calculated for lycopene, how far would this pigment travel if the solvent front moved 20 cm?

5. Tomato juice is red. What does this tell you about its pigment composition?

EXPERIMENT 4
Identification of alcohols and phenols

Background

Specific groups of atoms in an organic molecule can determine its physical and chemical properties. These groups are referred to as <u>functional groups</u>. Those hydrocarbons which contain the functions group -OH, the hydroxyl group, are called <u>alcohols</u>.

Alcohols are important commercially and include use as solvents, drugs and disinfectants. The most widely used alcohols are methanol or methyl alcohol, CH_3OH, ethanol or ethyl alcohol, CH_3CH_2OH, and 2-propanol or isopropyl alcohol, $(CH_3)_2CHOH$. Methyl alcohol is found in automotive products such as antifreeze and "dry gas." Ethyl alcohol is used as a solvent for drugs and chemicals, but is more popularly known for its effects as an alcoholic beverage. Isopropyl alcohol, also known as "rubbing alcohol," is an antiseptic.

Alcohols may be classified as either primary, secondary or tertiary:

$$R-CH_2-OH \qquad\qquad R-\underset{\underset{OH}{|}}{CH}-R' \qquad\qquad R-\underset{\underset{R''}{|}}{\overset{\overset{R'}{|}}{C}}-OH$$

Primary alcohol Secondary alcohol Tertiary alcohol

Note that the classification depends on the number of carbon containing groups, R (alkyl or aromatic), attached to the carbon bearing the hydroxyl group. Examples of each type are as follows:

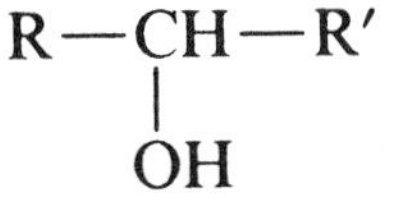

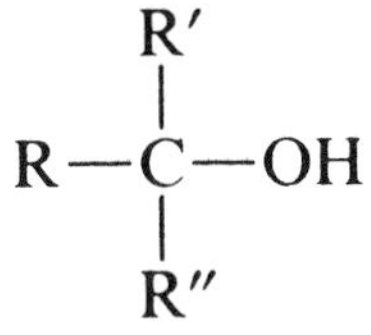

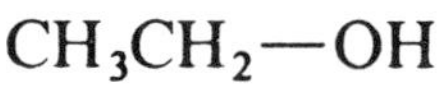

$$CH_3CH_2-OH \qquad\qquad CH_3-\underset{\underset{CH_3}{|}}{CH}-OH \qquad\qquad CH_3-\underset{\underset{CH_3}{|}}{\overset{\overset{CH_3}{|}}{C}}-OH$$

Ethyl alcohol Isopropyl alcohol <u>t</u>-Butyl alcohol
a primary alcohol a secondary alcohol a tertiary alcohol

Phenols bear a chose resemblance to alcohols structurally since the hydroxyl group is present. However, since the -OH group is bonded directly

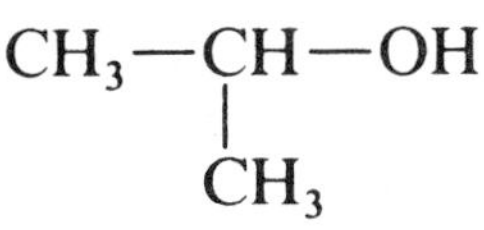

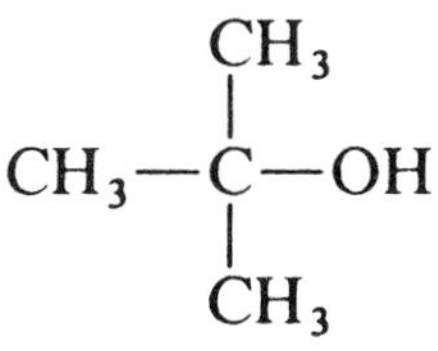

Table 4.1 Selected Alcohols and Phenols

Compound	Name and Use
CH_3OH	Methanol: solvent for paints, shellacs and varnishes
CH_3CH_2OH	Ethanol: alcoholic beverages; solvent for medicines, perfumes and varnishes
$CH_3-\underset{\displaystyle OH}{\underset{\vert}{CH}}-CH_3$	Isopropyl alcohol (2-propanol): rubbing alcohol; astringent; solvent for cosmetics, perfumes, and skin creams
$\underset{\displaystyle OH}{\underset{\vert}{CH_2}}-\underset{\displaystyle OH}{\underset{\vert}{CH_2}}$	Ethylene glycol: antifreeze
$\underset{\displaystyle OH}{\underset{\vert}{CH_2}}-\underset{\displaystyle OH}{\underset{\vert}{CH}}-\underset{\displaystyle OH}{\underset{\vert}{CH_2}}$	Glycerol (glycerin): sweetening agent; solvent for medicines; lubricant; moistening agent
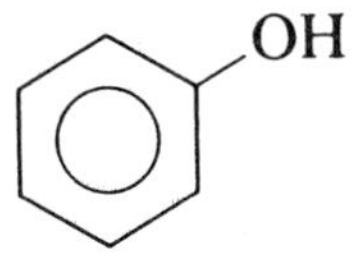	Phenol (carbolic acid): antiseptic; clean surgical and medical instruments
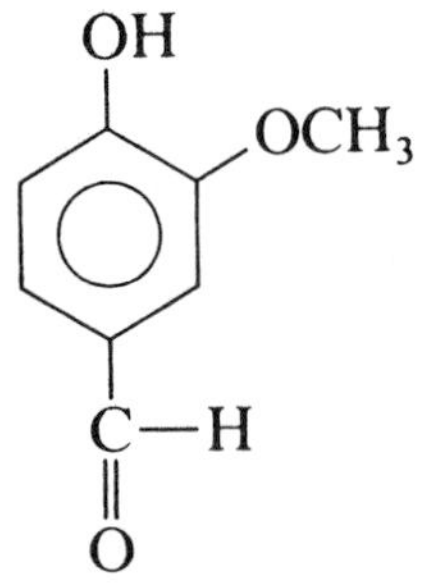	Vanillin: flavoring agent (vanilla)
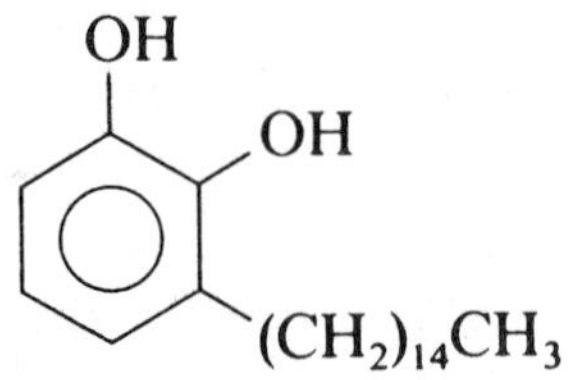	Tetrahydrourushiol: irritant in poison ivy

to a carbon that is part of an aromatic ring, the chemistry is quite different from that of alcohols. Phenols tend to be more acidic than alcohols; concentrated solutions of the compound phenol are quite toxic and can cause severe skin burns. Phenol derivatives are found in medicines; for example, thymol is used to kill fungi and hookworms. (Also see Table 4.1.)

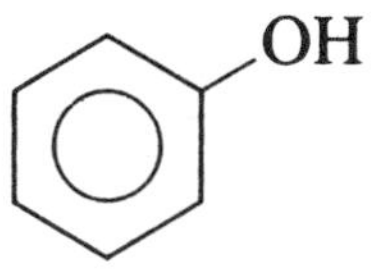

Phenol

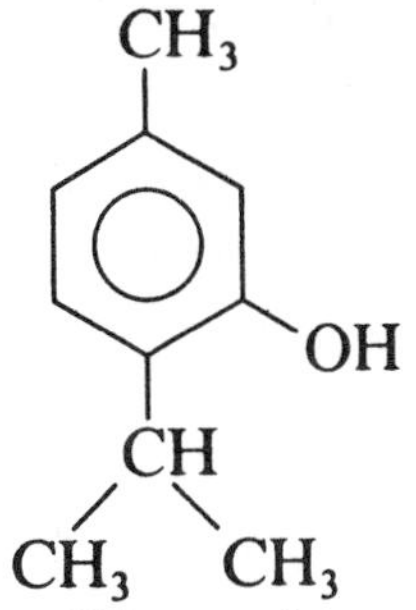

Thymol
(2-isopropyl-5-methylphenol)

In this laboratory, you will examine physical and chemical properties of representative alcohols and phenols. You will be able to compare the differences in chemical behavior between these compounds and use this information to identify an unknown.

Physical properties

Since the hydroxyl group is present in alcohols and phenols, these compounds are polar. The polarity of the hydroxyl group, coupled with its ability to form hydrogen bonds, enables alcohols and phenols to mix with water. Since these compounds also are carbon containing, they show additional solubility in many organic solvents, such as dichloromethane and diethyl ether.

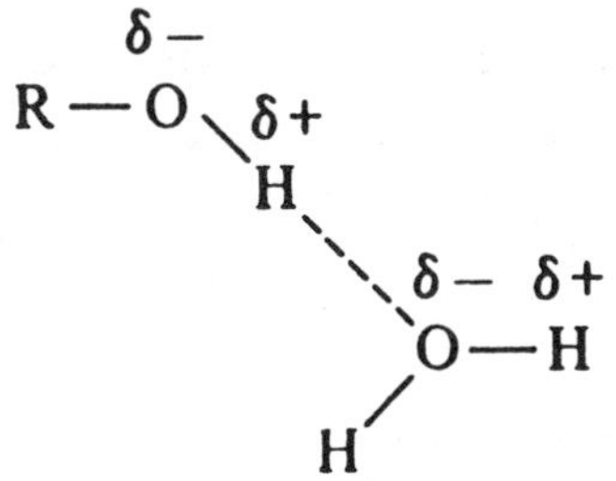

Hydrogen bonding of the hydroxyl group with water.

Chemical properties

The chemical behavior of the different classes of alcohols and of phenols can be used as a means of identification. Quick, simple tests that can be carried out in test tubes will be performed.

1. Lucas test. Alcohols react with Lucas reagent, a mixture of zinc chloride, $ZnCl_2$, in concentrated HCl, to give alkyl chlorides; these products, generally, are insoluble. Also, since this reaction takes place with different rates depending on the class of the alcohol, this test can be used to distinguish between primary, secondary and tertiary alcohols. Upon addition of this reagent a tertiary alcohol reacts rapidly and gives an insoluble white layer within 5 min. A secondary alcohol reacts slowly and gives the white layer within 20 or 30 min., usually with slight heating. A primary alcohol does not react. Any formation of a heterogeneous phase or appearance of an emulsion is a positive test.

$$\underset{\text{primary alcohol}}{CH_3CH_2{-}OH} + HCl + ZnCl_2 \longrightarrow \text{no reaction}$$

$$\underset{\text{secondary alcohol}}{(CH_3)_2CH{-}OH} + HCl + ZnCl_2 \longrightarrow \underset{\text{insoluble}}{(CH_3)_2CH{-}Cl\downarrow} + H_2O \text{ (20-30 min., heat)}$$

$$\underset{\text{tertiary alcohol}}{(CH_3)_3C{-}OH} + HCl + ZnCl_2 \longrightarrow \underset{\text{insoluble}}{(CH_3)_3C{-}Cl\downarrow} + H_2O \text{ (<5 min.)}$$

2. Chromic acid test. Chromic acid is a strong oxidizing agent and can oxidize primary and secondary alcohols. Thus, this test is able to distinguish primary and secondary alcohols from tertiary alcohols. Using acidified dichromate solution primary alcohols are oxidized to carboxylic acids; secondary alcohols are oxidized to ketones; tertiary alcohols are not oxidized. In the oxidation, the brown-red color of the chromic acid changes to a blue-green solution. Phenols are oxidized to nondescript brown tarry masses. (Aldehydes are also oxidized under these conditions, but ketones remain intact; see Experiment 5 for further discussion.)

$$3\ \underset{\text{primary alcohol}}{CH_3CH_2\text{-}OH} + 4\ \underset{\text{brown-red}}{H_2CrO_4} + 6\ H_2SO_4 \longrightarrow 3\ \underset{\text{carboxylic acid}}{CH_3\text{-}\overset{\overset{\displaystyle O}{||}}{C}\text{-}OH} + 2\ \underset{\text{blue-green}}{Cr_2(SO_4)_3} + 13\ H_2O$$

$$3\ \underset{\text{secondary alcohol}}{CH_3\text{-}\overset{\overset{\displaystyle OH}{|}}{CH}\text{-}CH_3} + 2\ \underset{\text{brown-red}}{H_2CrO_4} + 3\ H_2SO_4 \longrightarrow 3\ \underset{\text{ketone}}{CH_3\text{-}\overset{\overset{\displaystyle O}{||}}{C}\text{-}CH_3} + \underset{\text{blue-green}}{Cr_2(SO_4)_3} + 8\ H_2O$$

$$\underset{\text{tertiary alcohol}}{(CH_3)_3\text{-}OH} + H_2CrO_4 + H_2SO_4 \longrightarrow \text{no reaction}$$

3. Iodoform test. This test is more specific than the previous two tests. Only ethyl alcohol and alcohols with the part structure $CH_3CH(OH)$ react. These alcohols react with iodine in aqueous sodium hydroxide to give the yellow precipitate iodoform.

$$RCH(OH)CH_3 + 4\,I_2 + 6\,NaOH \longrightarrow RC(=O){-}O^-Na^+ + 5\,NaI + 5\,H_2O + HCI_{3(s)}$$

iodoform
yellow

Phenols also react under these conditions. With phenol, the yellow precipitate triiodophenol forms.

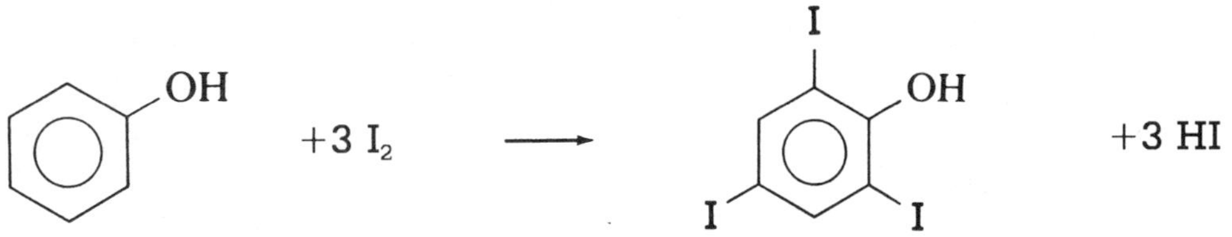

triiodophenol
yellow

4. Acidity of phenol. Phenol is also called carbolic acid. This indicates phenol is an acid and that it will react with base.

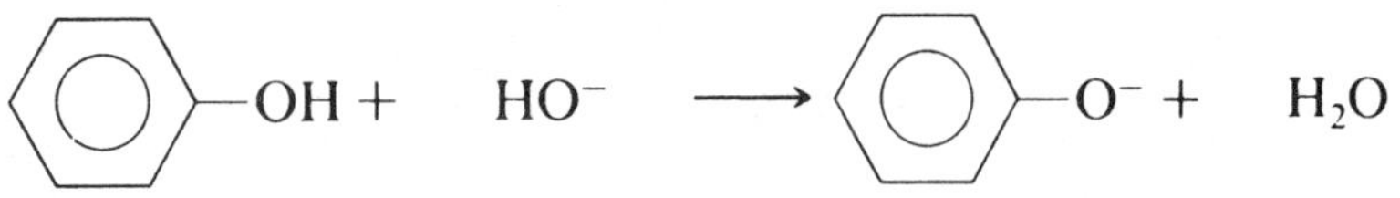

5. Ferric chloride test. Addition of aqueous ferric chloride to a phenol gives a colored solution. Depending on the structure of the phenol, the color can vary from green to purple.

light yellow violet color

Objectives

1. To learn characteristic chemical reactions of alcohols and phenols.
2. To use these chemical characteristics for identification of an organic compound.

Procedure

CAUTION! Chromic acid is very corrosive. Any spill should be immediately flushed with water. Phenol is toxic and will cause burns to skin. Any contact should be thoroughly washed with large quantities of water. Handle the solid only with a spatula or forceps. Dispose of reaction mixtures and excess reagents in proper containers as directed by your instructor.

Physical properties of alcohols and phenols

1. You will test the alcohols 1-butanol (a primary alcohol), 2-butanol (a secondary alcohol), t-butyl alcohol (a tertiary alcohol), and phenol; you will also have as an unknown one of these compounds (labeled A, B, C or D). As you run a test on a known, test the unknown at the same time for comparison. Note that the phenol will be provided as an aqueous solution.

2. Into separate test tubes (100 x 13 mm) labeled 1-butanol, 2-butanol, t-butyl alcohol and unknown place 10 drops of each sample; dilute by mixing with 3 mL of distilled water. Into a separate test tube place 2 mL of a prepared water solution of phenol. Are all the solutions homogeneous?

3. Test the pH of each of the aqueous solutions. Do the test by first dipping a clean glass rod into the solutions and then transferring a drop of liquid to pH paper. Use a broad indicator paper (e.g. pH range 1-12) and read the value of the pH by comparing the color to the chart on the dispenser. Record the results.

Chemical properties of alcohols and phenols

1. Lucas test. Place into separate clean, dry test tubes (100 x 13 mm) labeled 1-butanol, 2-butanol, t-butyl alcohol, phenol and unknown 10 drops of each sample. Add 40 drops of Lucas reagent; mix well by stoppering each test tube with a cork and shaking vigorously for a few seconds; remove the cork after shaking and allow each test tube to stand for 5 min. Look carefully for any cloudiness which may develop during this time period. If there is no cloudiness after 15 min., warm the test tubes that are clear for 15 min. in a 60°C water bath. Record your observations on the Report Sheet.

2. Chromic acid test. Place into separate clean, dry test tubes (100 x 13 mm), labeled as before, 5 drops of sample to be tested. To each test tube add 10 drops of reagent grade acetone and 2 drops chromic acid. Place the test tubes in a 60°C water bath for 5 min. Note the color of each solution. (Remember, the loss of the red-brown color and the formation of a blue-green color is a positive test.) Record your observations on the Report Sheet.

3. Iodoform test. Place into separate clean, dry test tubes (150 x 18 mm), labeled as before, 10 drops of sample to be tested. Add to each test tube dropwise, with shaking, 25 drops of 6 M NaOH. The mixture is warmed in a 60°C water bath, and the prepared solution of I_2-KI test reagent is added dropwise, with shaking, until the solution becomes brown (approx. 35 drops). Add 6 M NaOH until the solution becomes colorless. Keep the test tubes in the warm water bath for 5 min. Remove the test tubes from the water, let cool and look for a light yellow precipitate. Record your observations on the Report Sheet.

4. Ferric chloride test. Place into separate clean, dry test tubes (100 x 13 mm), labeled as before, 20 drops of sample to be tested. Add 5 drops of ferric chloride solution to each. Note any color changes in each solution. (Remember, a purple color indicates the presence of a phenol.) Record your observations on the Report Sheet.

Chemicals and Equipment

1. Acetone (reagent grade)
2. 1-Butanol
3. 2-Butanol

4. t-Butyl alcohol
5. Aqueous phenol
6. Lucas reagent
7. Chromic acid solution
8. Ferric chloride solution
9. I_2-KI solution
10. pH paper
11. Corks
12. Hot plate
13. Unknown

NAME SECTION DATE

PARTNER GRADE

Experiment 4

PRE-LAB QUESTIONS

1. Give an example for the functional groups given below:

   a. a primary alcohol

   b. a secondary alcohol

   c. a tertiary alcohol

2. The compound below, tetrahydrourushiol, is an example of a compound which contains the phenolic functional group. Circle the phenol part of the molecule.

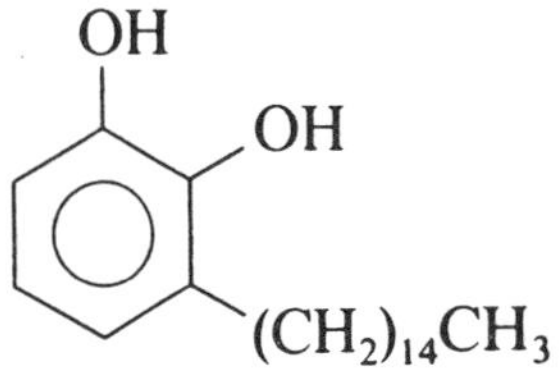

3. What is present in alcohols that makes these compounds more soluble in water than hydrocarbons?

NAME ______________________ SECTION __________ DATE __________

PARTNER ______________________ GRADE ______________________

Experiment 4

REPORT SHEET

Test	1-Butanol	2-Butanol	t-Butyl Alcohol	Phenol	Unknown
1. pH					
2. Lucas					
3. Chromic Acid					
4. Iodoform					
5. Ferric Chloride					

Identity of unknown:

Unknown No.____________. The unknown compound is________________.

POST-LAB QUESTIONS

1. Write the structures for the following compounds:

 a. 1-butanol

 b. 2-butanol

 c. t-butyl alcohol

2. Write the structure of the major organic product expected from each of the following reactions. If no reaction is expected write "No Reaction."

 a.
 $$\text{C}_6\text{H}_5\text{—OH} + \text{NaOH} \longrightarrow$$

 b.
 $$\text{CH}_3\text{—CH—CH}_2\text{CH}_3 + \text{H}_2\text{CrO}_4 \xrightarrow{\text{H}_2\text{SO}_4}$$
 $$\quad\quad\quad | $$
 $$\quad\quad\quad \text{OH}$$

 c.
 $$\text{CH}_3\text{CH}_2\text{—} \overset{\overset{\displaystyle \text{CH}_3}{|}}{\underset{\underset{\displaystyle \text{CH}_3}{|}}{\text{C}}} \text{—OH} + \text{H}_2\text{CrO}_4 \xrightarrow{\text{H}_2\text{SO}_4}$$

 d.
 $$\text{CH}_3\text{—} \overset{\overset{\displaystyle \text{CH}_3}{|}}{\underset{\underset{\displaystyle \text{CH}_3}{|}}{\text{C}}} \text{—OH} + \text{HCl} \xrightarrow{\text{ZnCl}_2}$$

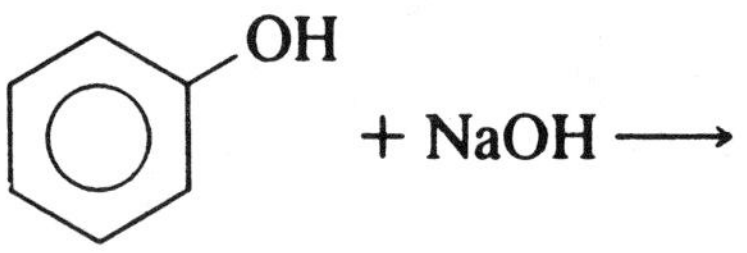

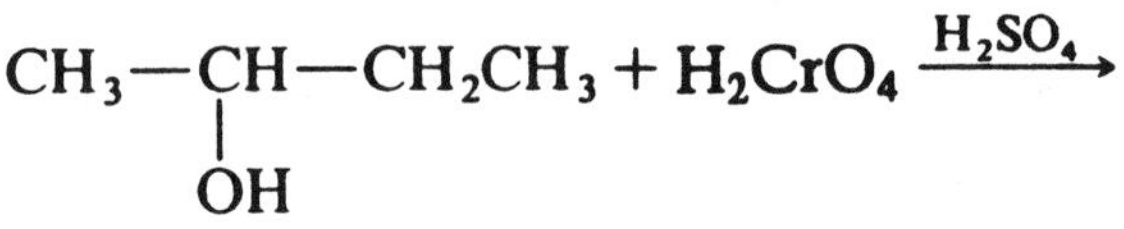

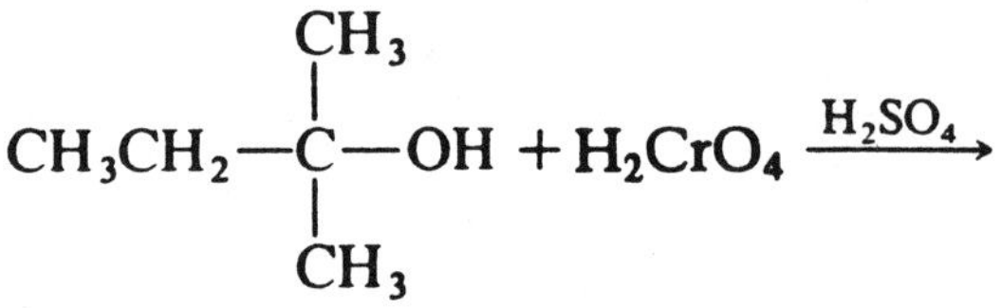

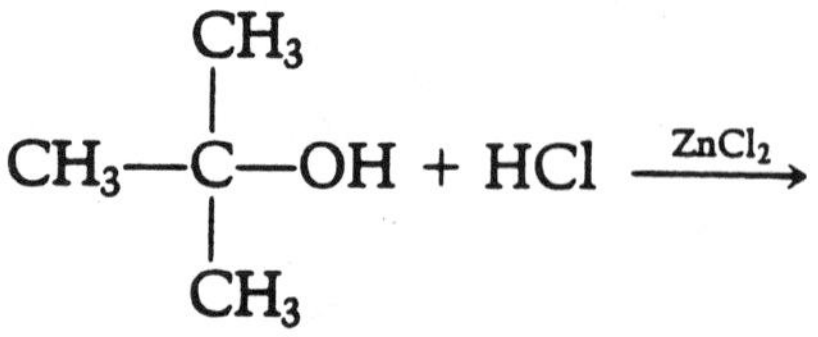

EXPERIMENT 5

Identification of aldehydes and ketones

Background

Aldehydes and ketones are representative of compounds which possess the carbonyl group:

$$\gt C = O \qquad \text{The carbonyl group.}$$

Aldehydes have at least one hydrogen attached to the carbonyl carbon; in ketones, no hydrogens are directly attached to the carbonyl carbon, only carbon containing R-groups:

$$R{-}\overset{\overset{\displaystyle O}{\|}}{C}{-}H \qquad\qquad R{-}\overset{\overset{\displaystyle O}{\|}}{C}{-}R'$$

Aldehyde　　　Ketone　　　(R is alkyl or aromatic)

Aldehydes and ketones of low molecular weight have commercial importance. Many others occur naturally. Table 5.1 has some representative examples.

Table 5.1 Representative Aldehydes and Ketones

Compound		Source and Use
$H\overset{\overset{O}{\|}}{C}H$	Formaldehyde	Oxidation of methanol; plastics; preservative
$CH_3\overset{\overset{O}{\|}}{C}CH_3$	Acetone	Oxidation of isopropyl alcohol; solvent
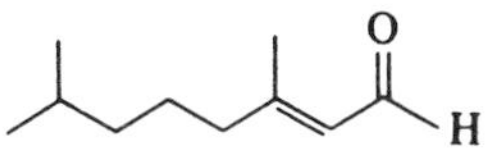	Citral	Lemon grass oil; fragrance
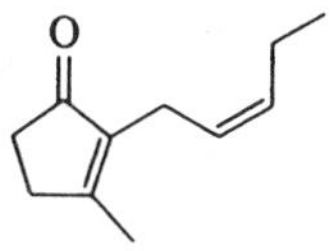	Jasmone	Oil of jasmine; fragrance

In this experiment you will investigate the chemical properties of representative aldehydes and ketones.

1. Chromic acid test. Aldehydes are oxidized to carboxylic acids by chromic acid; ketones are not oxidized. A positive test is the formation of a blue-green solution from the brown-red color of chromic acid.

$$3R-\overset{O}{\overset{\|}{C}}-H + 2H_2CrO_4 + 3H_2SO_4 \rightarrow 3R-\overset{O}{\overset{\|}{C}}-OH + Cr_2(SO_4)_3 + 5H_2O$$

aldehyde   brown-red   blue-green

$$R_2C{=}O \xrightarrow[H_2SO_4]{H_2CrO_4} \text{no reaction}$$

ketone

2. Tollens' test. Most aldehydes reduce Tollens' reagent (ammonia and silver nitrate) to give a precipitate of silver metal. The free silver forms a silver mirror on the sides of the test tube. (This test is sometimes referred to as the "silver mirror" test.) The aldehyde is oxidized to a carboxylic acid.

$$R\overset{O}{\overset{\|}{C}}-H + 2\ Ag(NH_3)_2OH \longrightarrow 2\ Ag_{(s)} + R\overset{O}{\overset{\|}{C}}O^-NH_4^+ + H_2O + 3\ NH_3$$

aldehyde   silver mirror

3. Benedict's test. Aliphatic aldehydes are oxidized to carboxylic acids with Benedict's reagent (a citrate complex of Cu(II) ion in a base solution). The Cu(II) ion is reduced to Cu(I) ion and its oxide, Cu_2O, precipitates as a brick-red solid. Aromatic aldehydes and ketones do not react.

$$R-CHO + 2Cu(OH)_2 + NaOH \rightarrow RCOO^-\ Na^+ + Cu_2O(s) + 3H_2O$$

aldehyde   blue   red

$$R_2C{=}O + 2Cu(OH)_2 + NaOH \rightarrow \text{no reaction}$$

ketone

4. Iodoform test. Methyl ketones give the yellow precipitate iodoform when reacted with iodine in aqueous sodium hydroxide.

$$\overset{\displaystyle O}{\overset{\|}{RC}}{-}CH_3 + 3\,I_2 + 4\,NaOH \longrightarrow 3\,NaI + 3\,H_2O + \overset{\displaystyle O}{\overset{\|}{RC}}{-}O^-Na^+ + HCI_{3(s)}$$

methyl ketone — iodoform, yellow

5. 2,4-Dinitrophenylhydrazine test. All aldehydes and ketones give an immediate precipitate with 2,4-dinitrophenylhydrazine reagent. This reaction is general for both these functional groups. The color of the precipitate varies from yellow to red. (Note that alcohols do not give this test.)

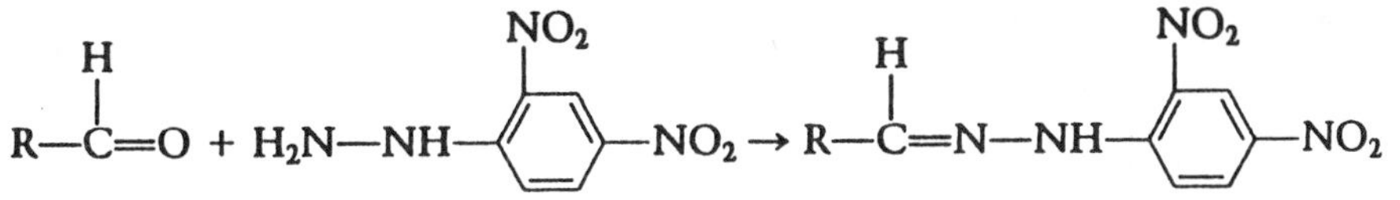

aldehyde — yellow to red

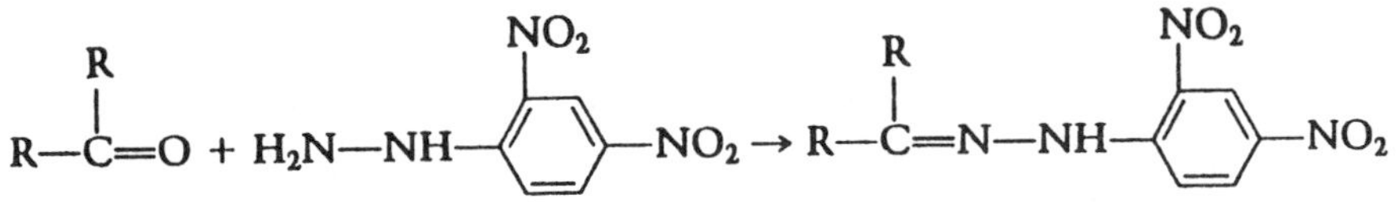

ketone — yellow to red

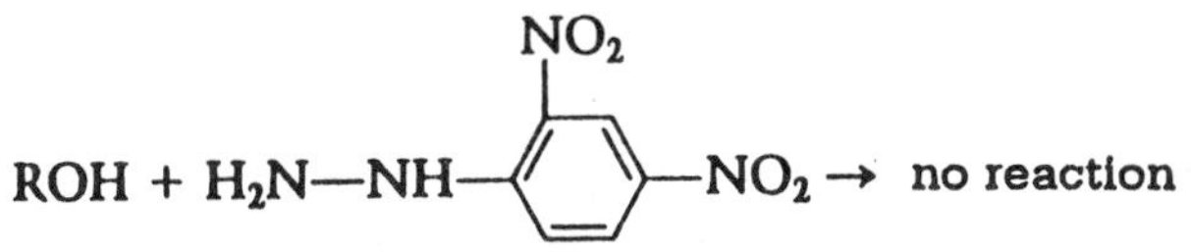

Table 5.2 Summary of Classification Tests

Compound	Test
Aldehydes and Ketones	2,4-Dinitrophenylhydrazine
Aldehydes	Chromic acid
	Tollens' reagent
Aliphatic Aldehydes	Benedict's reagent
Methyl Ketones	Iodoform

Objectives

1. To learn the chemical characteristics of aldehydes and ketones.
2. To use these chemical characteristics in simple tests to distinguish between examples of aldehydes and ketones.

Procedure

1. Classification tests are to be carried out on four known compounds and one unknown. Any one test should be carried out on all five samples at the same time for comparison. Label test tubes as shown in Table 5.3.

Table 5.3 Labeling Test Tubes

Test Tube No.	Compound
1	Isovaleraldehyde (an aliphatic aldehyde)
2	Benzaldehyde (an aromatic aldehyde)
3	Cyclohexanone (a ketone)
4	Acetone (a methyl ketone)
5	Unknown

CAUTION! Chromic acid is corrosive. Handle with care and wash promptly any spill.

2. Chromic acid test. Place 5 drops of each substance into separate, labeled test tubes (100 x 33 mm). Dissolve each compound in 20 drops of reagent grade acetone (to serve as solvent); then add to each test tube 4 drops of chromic acid reagent, one drop at a time, with shaking. Cork and shake each test tube to insure thorough mixing. Remove the corks and let stand for 10 min. Aliphatic aldehydes should show a change within a minute; aromatic aldehydes take longer. Note the approximate time for any change in color or formation of a precipitate on the Report Sheet.

3. Tollens' test.

CAUTION! The reagent must be freshly prepared before it is to be used and any excess disposed of immediately after use. All residues should be discarded in appropriate waste containers or flushed in the sink with large quantities of water. Do not store Tollens' reagent since it is explosive when dry.

Enough reagent for your use can be prepared in a 25-mL Erlenmeyer flask by mixing 5 mL of Tollens' solution A with 5 mL of Tollens' solution B. To the silver oxide precipitate which forms, add dropwise, with shaking, 10% ammonia solution until the brown precipitate just dissolves. Avoid an excess of ammonia.
Place 5 drops of each substance into separately labeled clean, dry test tubes (100 x 13 mm). Dissolve the compound in bis(2-ethoxyethyl)ether by adding this solvent dropwise until a homogeneous solution is obtained. Then add 2 mL (40 drops) of the prepared Tollens' reagent, cork the test tube and shake vigorously. Remove the cork and place the test tube in a 60°C water bath for 5 min. Remove the test tubes from the water and look for a silver mirror. Record your results on the Report Sheet.

4. Benedict's test. Prepare a hot water bath (above 90°C) for this part. Place 10 drops of each substance into separately labeled clean, dry test tubes (100 x 13 mm). Add 2 mL (40 drops) of Benedict's reagent to each test tube. Cork each test tube and shake vigorously. Remove the corks and then place the test tubes in the warm water bath and heat for 10 min. Remove the test tubes from the water bath, let cool and look for an orange to brick-red precipitate. Record your observations on the Report Sheet.

5. Iodoform test. Place 10 drops of each substance into separately labeled clean, dry test tubes (150 x 18 mm). Add to each test tube, dropwise, with shaking, 25 drops of 6 M NaOH. The mixture is warmed in a water bath (60°C), and the prepared solution of I_2-KI test reagent is added dropwise, with shaking, until the solution becomes brown (approx. 35 drops). Add 6 M NaOH until the solution becomes colorless. Keep the test tubes in the warm water bath for 5 min. Remove the test tubes from the water, let cool and look for a light yellow precipitate. Record your observations on the Report Sheet.

6. 2,4-Dinitrophenylhydrazine test. Place 5 drops of each substance into separately labeled clean, dry test tubes (100 x 13 mm) and add 20 drops of the 2,4-dinitrophenylhydrazine reagent to each. Cork each test tube and shake the mixture vigorously. If no precipitate forms remove the cork from the test tube and heat for 5 min. in a warm water bath (60°C); cool. Record your observations on the Report Sheet.

7. Based on the observations you recorded on the Report Sheet, and by comparing the results of your unknown to the knowns, identity your unknown.

Chemicals and Equipment

1. Acetone (reagent grade)
2. Isovaleraldehyde
3. Benzaldehyde
4. Cyclohexanone
5. Bis(2-ethoxyethyl) ether
6. Chromic acid reagent
7. Tollens' reagent (solution A and solution B)
8. 10% ammonia solution
9. Benedict's reagent
10. 6M NaOH
11. I_2-KI test solution
12. 2,4-Dinitrophenylhydrazine reagent
13. Hot plate

NAME SECTION DATE

PARTNER GRADE

Experiment 5

PRE-LAB QUESTIONS

1. Write a general structure for the following functional groups:

   a. an aldehyde

   b. a ketone

   c. a methyl ketone

   d. an aromatic aldehyde

2. What reagent gives a characteristic test for a methyl ketone?

3. Tollens' reagent will test for which functional group?

NAME ______________________ SECTION __________ DATE __________

PARTNER ______________________ GRADE __________

Experiment 5

REPORT SHEET

Test	Isovaler-aldehyde	Benz-aldehyde	Cyclo-hexanone	Acetone	Unknown
Chromic acid					
Tollens'					
Benedict's					
Iodoform					
2,4-Dinitro-phenyl-hydrazine					

Identity of unknown:

Unknown No.________. The unknown compound is ______________________.

POST-LAB QUESTIONS

1. A compound of molecular formula $C_5H_{10}O$ forms a yellow precipitate with 2,4-dinitrophenylhydrazine reagent and a yellow precipitate with reagents for the iodoform test. Draw a structural formula for the compound that fits these tests.

2. What kind of results do you expect to see when the following compounds are mixed together with the given test solution?

a. 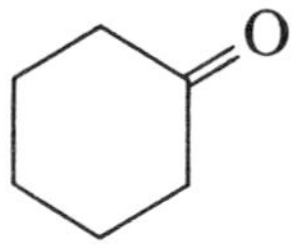 with 2,4-dinitrophenylhydrazine

b. 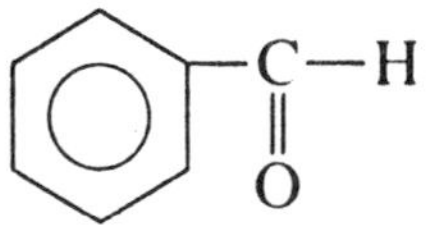 with chromic acid

c. 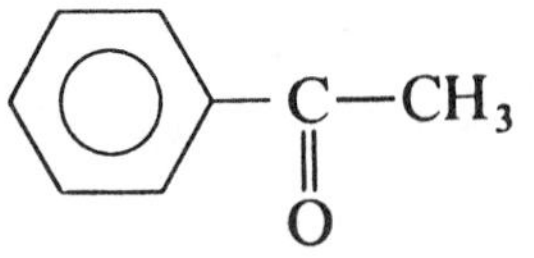 with I_2-KI reagent

d. $CH_3CH_2-\underset{\underset{O}{\|}}{C}-H$ Tollens' reagent

EXPERIMENT 6
Carboxylic acids and esters

Background

Carboxylic acids are structurally like aldehydes and ketones in that they contain the carbonyl group. However, an important difference is that carboxylic acids contain a hydroxyl group attached to the carbonyl carbon.

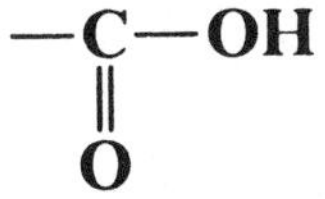

$$-\underset{\underset{O}{\|}}{C}-OH \qquad \text{The carboxylic acid group.}$$

This combination gives the group its most important characteristic; it behaves as an acid.

As a family carboxylic acids are weak acids which ionize only slightly in water. As aqueous solutions typical carboxylic acids ionize to the extent of only one percent or less.

$$R-\underset{\underset{O}{\|}}{C}-OH + H_2O \rightleftharpoons R-\underset{\underset{O}{\|}}{C}-O^- + H_3O^+$$

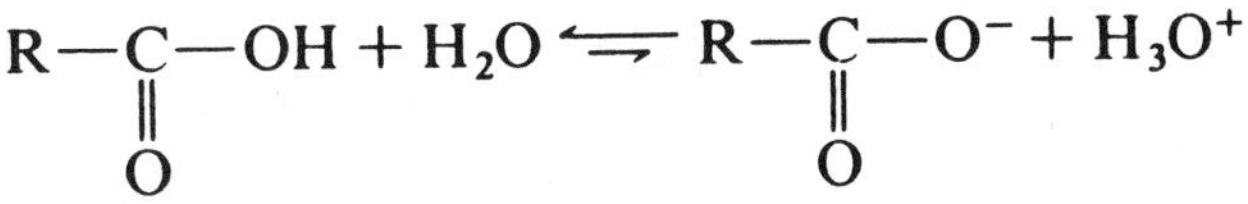

At equilibrium most of the acid is present as unionized molecules. Dissociation constants, K_a, of carboxylic acids, where R is an alkyl group, are 10^{-5} or less. Water solubility depends to a large extent on the size of the R-group. Only a few low molecular weight acids (up to four carbons) are very soluble in water.

Although carboxylic acids are weak, they are capable of reacting with bases stronger than water. Thus, while benzoic acid shows limited water solubility, it reacts with sodium hydroxide to form the soluble salt sodium benzoate.

$$\langle \bigcirc \rangle\!-COOH + NaOH \rightarrow \langle \bigcirc \rangle\!-COO^- Na^+ + H_2O$$

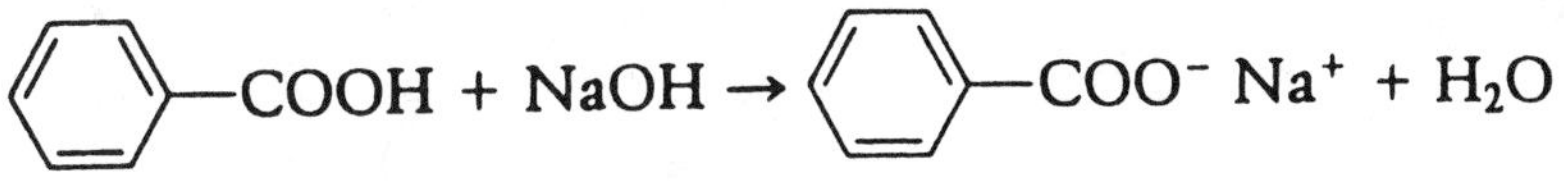

Insoluble Soluble

Sodium carbonate, Na_2CO_3, and sodium bicarbonate, $NaHCO_3$, solutions can neutralize carboxylic acids, also.

The combination of a carboxylic acid and an alcohol gives an ester; water is eliminated. Ester formation is an equilibrium process, catalyzed by an acid catalyst.

$$CH_3CH_2CH_2\overset{\overset{\displaystyle O}{\|}}{C}{-}OH + CH_3CH_2OH \overset{H^+}{\rightleftharpoons} H_2O + CH_3CH_2CH_2\overset{\overset{\displaystyle O}{\|}}{C}{-}OCH_2CH_3$$

Butyric acid | Ethyl alcohol | Ethyl butyrate (Ester)

Esterification →

← Hydrolysis

The reaction typically gives 60% to 70% of the maximum yield. The reaction is a reversible process. An ester reacting with water, giving the carboxylic acid and alcohol, is called <u>hydrolysis</u>; it is acid catalyzed. The base-promoted decomposition of esters yields an alcohol and a salt of the carboxylic acid; this process is called <u>saponification</u>.

$$CH_3CH_2CH_2\overset{\overset{\displaystyle O}{\|}}{C}{-}OCH_2CH_3 + NaOH \longrightarrow CH_3CH_2CH_2\overset{\overset{\displaystyle O}{\|}}{C}{-}O^-Na^+ + CH_3CH_2OH$$

Saponification

A distinctive difference between carboxylic acids and esters is in their characteristic odors. Carboxylic acids are noted for their sour, disagreeable odors. On the other hand, esters have sweet and pleasant odors often associated with fruits; these compounds are used in the food industry as flavoring agents. For example, while butyric acid gives rancid butter its putrid odor, the ester, ethyl butyrate, is used as the artificial flavoring agent of pineapple. Only those carboxylic acids of low molecular weight have odor at room temperature. Higher molecular weight carboxylic acids form strong hydrogen bonds, are solid and have a low vapor pressure. Thus, few molecules reach our nose. Esters, however, do not form hydrogen bonds among themselves; they are liquid at room temperature, even when the molecular weight is high. Thus, they have high vapor pressure and many molecules can reach our nose providing odor.

Objectives

1. To study the physical properties of carboxylic acids: solubility, acidity, aroma.
2. To prepare a variety of esters and note their odors.
3. To demonstrate saponification.

Procedure

Carboxylic acids and their salts

Characteristics of Acetic Acid

1. Place into a clean, dry test tube (100 x 13 mm) 2 mL of water and 10 drops of glacial acetic acid. Note its odor. Of what does it remind you?

2. Take a glass rod and dip it into the solution. Using wide range indicator paper (pH 1-12), test the pH of the solution. Determine the value of the pH by comparing the color of the paper with the chart on the dispenser.

3. Now add 2 mL of 2 M NaOH to the solution. Cork the test tube and shake. Remove the cork and determine the pH of the solution as before; if not basic, continue to add more base, dropwise, until the solution is basic. Note the odor and compare the odor to the solution before the addition of base.

4. By dropwise addition of 3 M HCl, carefully reacidify the solution from 3 (above); test the solution as before with pH paper until the solution tests acid. Does the original odor return?

Characteristics of Benzoic Acid

1. Your instructor will weigh out 0.1 g of benzoic acid for sample size comparison. With your microspatula take some sample equivalent to the preweighed sample (an exact quantity is not important here). Add the solid to a test tube (100 x 13 mm) along with 2 mL of water. Is there any odor? Shake the mixture. How soluble is the benzoic acid?

2. Now add 1 mL of 2 M NaOH to the solution from 1 (above), cork and shake. What happens to the solid benzoic acid? Is there any odor?

3. By dropwise addition of 3 M HCl, carefully reacidify the solution from 2 (above); test as before with pH paper until acidic. As the solution becomes acidic what do you observe?

Esterification

1. Into five clean, dry test tubes (100 x 13 mm) add 10 drops of liquid carboxylic acid or 0.1 g of solid carboxylic acid and 10 drops of alcohol according to the scheme in Table 6.1. Note the odor of each reactant.

Table 6.1 Acids and Alcohols

Test Tube No.	Carboxylic Acid	Alcohol
1	Formic	Isobutyl
2	Acetic	Benzyl
3	Acetic	Isoamyl
4	Acetic	Ethyl
5	Salicylic	Methyl

2. Add 5 drops of concentrated sulfuric acid to each test tube and mix the contents thoroughly.

CAUTION! Sulfuric acid causes severe burns. Flush any spill with lots of water.

3. Place the test tubes in a warm water bath at 60°C for 15 min. Remove the test tubes from the water bath, cool and add 2 mL of water to each. Note that there is a layer on top of the water in each test tube. With a Pasteur pipet, take a few drops from this top layer and place on a watch glass. Note the odor. Match the ester from each test tube with one of the following odors: banana, peach, raspberry, nail polish remover, wintergreen.

Saponification

This part of the experiment can be done while the esterification reactions are being heated.

1. Place into a test tube (150 x 18 mm) 10 drops of methyl salicylate and 5 mL of 6 M NaOH. Heat the contents in a boiling water bath for 30 min. What has happened to the ester layer?

2. Cool the test tube to room temperature by placing it in a cold stream of water. What has happened to the odor of the ester?

3. Carefully add 6 M HCl to the solution, 1 mL at a time, until the solution is acidic. After each addition, mix the contents and test the solution with litmus. When the solution is acidic what do you observe? What is the name of the compound formed?

Chemicals and Equipment

1. Glacial acetic acid
2. Benzoic acid
3. Formic acid
4. Salicylic acid
5. Benzyl alcohol
6. Ethyl alcohol
7. Isobutyl alcohol
8. Isoamyl alcohol
9. Methyl alcohol
10. Methyl salicylate
11. 3 M HCl
12. 6 M HCl
13. 2 M NaOH
14. 6 M NaOH
15. Concentrated H_2SO_4
16. pH paper (broad range pH 1-12)
17. Litmus paper
18. Pasteur pipet
19. Hot plate

NAME SECTION DATE

PARTNER GRADE

Experiment 6

PRE-LAB QUESTIONS

1. Write the structures of the following carboxylic acids:

   a. salicylic acid

   b. benzoic acid

   c. acetic acid

2. Write the products from formic acid and sodium hydroxide.

3. n-Propyl acetate has the odor of pears. What alcohol and carboxylic acid would you use to synthesize this ester?

4. Esters can be decomposed by either hydrolysis or saponification. Explain the difference between the methods in terms of conditions and end products.

NAME ____________________ SECTION __________ DATE __________

PARTNER ____________________ GRADE __________

Experiment 6

REPORT SHEET

Carboxylic acids and their salts

Characteristics of Acetic Acid

Property	Water Solution	NaOH Solution	HCl Solution
Odor			
Solubility			
pH			

Characteristics of Benzoic Acid

Property	Water Solution	NaOH Solution	HCl Solution
Odor			
Solubility			
pH			

Esterification

Test Tube	Acid	Odor	Alcohol	Odor	Ester	Odor
1	Formic		Isobutyl			
2	Acetic		Benzyl			
3	Acetic		Isoamyl			
4	Acetic		Ethyl			
5	Salicylic		Methyl			

Saponification

1. Write the chemical equation for the saponification of methyl salicylate.

2. What has happened to the ester layer?

3. What has happened to the odor of the ester?

4. What forms on reacidification of the solution? Name the compound.

POST-LAB QUESTIONS

1. Explain why acetic acid has an odor, but benzoic acid does not?

2. Write equations for each of the five esterification reactions.

3. Butyric acid has a putrid odor (like rancid butter). Suppose you got some on your hands. How could you rid your hands of the odor? (Remember, butyric acid has marginal solubility in water.)

EXPERIMENT 7

Amines and amides

Background

Two classes of organic compounds which contain nitrogen are amines and amides. Amines behave as organic bases and may be considered as derivatives of ammonia. Amides are compounds which have a carbonyl group connected to a nitrogen atom. In this experiment you will learn about the physical and chemical properties of some members of the amine and amide families.

If the hydrogens of ammonia are replaced by alkyl or aryl groups, amines result. Depending on the number of organic groups attached to nitrogen, amines are classified as either primary (one group), secondary (two groups), or tertiary (three groups) (Table 7.1).

Table 7.1 Types of Amines

	Primary Amines	**Secondary Amines**	**Tertiary Amines**
NH_3 Ammonia	CH_3NH_2 Methylamine	$(CH_3)_2N$ Dimethylamine	$(CH_3)_3N$ Trimethylamine
	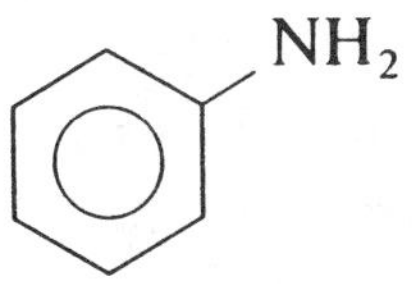 Aniline	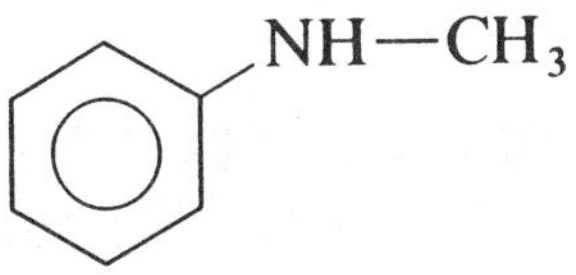 N-Methylaniline	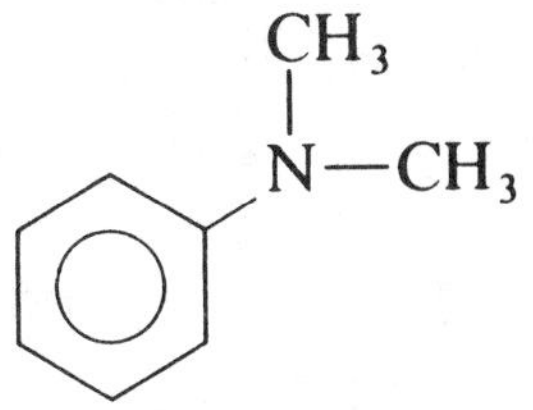 N,N-Dimethylaniline

There are a number of similarities between ammonia and amines that carry beyond the structural. Consider odor. The smell of amines resembles that of ammonia, but are not as sharp. However, amines can be quite pungent. Anyone handling or working with raw fish knows how strong the amine odor can be since raw fish contains low-molecular weight amines such as dimethylamine and trimethylamine. Other amines associated with decaying flesh have names suggestive of their odors: putrescine and cadaverine.

$NH_2CH_2CH_2CH_2CH_2NH_2$	$NH_2CH_2CH_2CH_2CH_2CH_2NH_2$
Putrescine	Cadaverine
(1,4-Diaminobutane)	(1,5-Diaminopentane)

The solubility of low molecular weight amines in water is high. In general, if the total number of carbons attached to nitrogen is six or less, the amine is water soluble; amines with a carbon content greater than six are water insoluble. However, all amines are soluble in organic solvents such as diethyl ether or methylene chloride.

Since amines are organic bases, water solutions show weakly basic properties. If the basicity of aliphatic amines and aromatic amines are compared to ammonia, aliphatic amines are stronger than ammonia, while aromatic amines are weaker. Amines characteristically react with acids to form ammonium salts.

$$\underset{\text{Amine}}{RNH_2} + HCl \longrightarrow \underset{\text{Ammonium Salt}}{RNH_3^+Cl^-}$$

If an amine is insoluble, reaction with an acid produces a water soluble salt. Since ammonium salts are water soluble, many drugs containing amines are prepared as ammonium salts. After working with fish in the kitchen, a convenient way to rid one's hands of fish odor is to rub a freshly cut lemon over the hands. The citric acid found in the lemon reacts with the amines found on the fish; a salt forms which can be easily rinsed away with water.

Amides are carboxylic acid derivatives. The amide group is recognized by the nitrogen connected to the carbonyl group. Amides are neutral compounds.

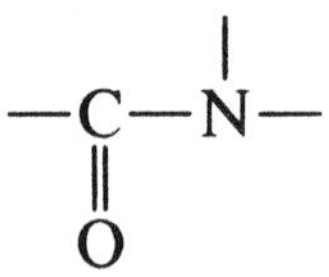

Amide group

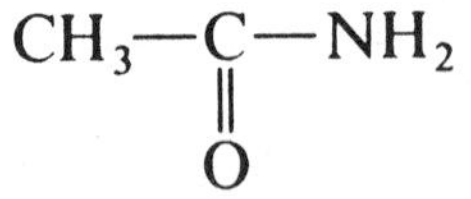

Acetamide

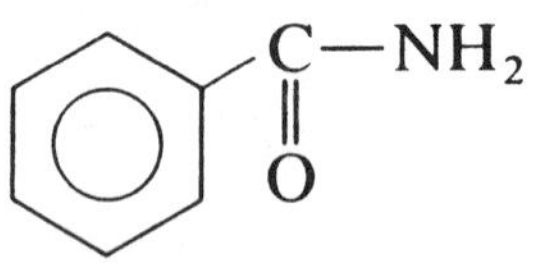

Benzamide

Under suitable conditions amide formation can take place between an amine and a carboxylic acid, an acyl halide or an acid anhydride. Along with ammonia, primary and secondary amines yield amides with carboxylic acids or derivatives. Table 7.2 relates the nitrogen base with the amide class

(based on the number of alkyl or aryl groups on the nitrogen of the amide.)

$$CH_3NH_2 + CH_3COOH \longrightarrow CH_3COO^-(CH_3NH_3^+) \xrightarrow{\Delta} CH_3CONHCH_3 + H_2O$$

$$CH_3NH_2 + CH_3COCl \longrightarrow CH_3CONHCH_3 + HCl$$

Table 7.2 Classes of Amides

Nitrogen Base	**Amide ($-CO-\overset{|}{N}-$)**
Ammonia	Primary amide (no R groups)
Primary Amine	Secondary amide (one R group)
Secondary Amine	Tertiary amide (two R groups)

Hydrolysis of amides can take place in either acid or base. Primary amides hydrolyze in acid to ammonium salts and carboxylic acids. Neutralization of the acid and ammonium salts releases ammonia which can be detected by odor or by litmus.

$$R\text{-}\overset{\overset{\displaystyle O}{\|}}{C}\text{-}NH_2 + HCl + H_2O \longrightarrow R\text{-}\overset{\overset{\displaystyle O}{\|}}{C}\text{-}OH + NH_4Cl$$

$$NH_4Cl + NaOH \longrightarrow NH_3 + NaCl + H_2O$$

Secondary and tertiary amides would release the corresponding alkyl ammonium salts which, when neutralized, would yield the amine.

In base primary amides hydrolyze to carboxylic acid salts and ammonia. The presence of ammonia (or amine from corresponding amides) can be detected similarly by odor or litmus. The carboxylic acid would be generated by neutralization with acid.

$$R\text{-}\overset{\overset{\displaystyle O}{\|}}{C}\text{-}NH_2 + NaOH \longrightarrow R\text{-}\overset{\overset{\displaystyle O}{\|}}{C}\text{-}O^-Na^+ + NH_3$$

$$R\text{-}\overset{\overset{\displaystyle O}{\|}}{C}\text{-}O^-Na^+ + HCl \longrightarrow R\text{-}\overset{\overset{\displaystyle O}{\|}}{C}\text{-}OH + NaCl$$

Objectives

1. To show some physical and chemical properties of amines and amides.
2. To demonstrate the hydrolysis of amides.

Procedure

CAUTION! Amines are toxic chemicals. Avoid excessive inhaling of the vapors and direct skin contact. Wash any amine spill with large quantities of water.

Properties of amines

1. Place 5 drops of liquid or 0.1 g of solid from the compounds listed in the table below into labeled clean, dry test tubes (100 x 13 mm).

Test Tube No.	Nitrogen Compound
1	6 M NH_3
2	Triethylamine
3	Aniline
4	N,N-Dimethylaniline
5	Acetamide

2. Carefully note the odors of each compound. Do not inhale deeply. Merely wave your hand across the mouth of the test tube toward your nose in order to note the odor. Record your observations on the Report Sheet.

3. Add 2 mL of distilled water to each of the labeled test tubes. Stopper with a cork and mix thoroughly by shaking. Note on the Report Sheet whether the amines are soluble or insoluble.

4. Take a glass rod and test each solution for its pH. Carefully dip one end of the glass rod into a solution and touch a piece of pH paper. Between each test, be sure to clean and dry the glass rod. Record the pH by comparing the color of the paper with the chart on the dispenser.

5. Carefully add 2 mL of 6 M HCl to each test tube. Stopper each tube with a cork and mix thoroughly by shaking. Compare the odor and solubility of this solution to previous observations.

6. Place 5 drops of liquid or 0.1 g of solid from the compounds listed in the table into labeled clean, dry test tubes (100 x 13 mm). Add 2 mL of diethyl ether (ether) to each test tube. Stopper with a cork and mix thoroughly by shaking. Record the observed solubilities.

7. Carefully place on a watch glass, side-by-side, without touching, a drop of triethylamine and a drop of concentrated HCl. Record your observations.

Hydrolysis of acetamide

1. Dissolve 0.5 g of acetamide in 5 mL of 6 M H_2SO_4 in a large test tube (150 x 18 mm). Heat the solution in a boiling water bath for 5 min.

2. Hold a small strip of moist pH paper over the mouth of the test tube; note any changes in color; record the pH reading. Remove the test tube from the water bath holding it in a test tube holder. Carefully note any odor.

3. Cool the test tube by running cold water over the outside of the tube. Now carefully add, dropwise, with shaking, 6 M NaOH to the cool solution until basic. (You will need more than 7 mL of base.) Hold a piece of moist pH paper over the mouth. Record the pH reading. Carefully note any odor.

Chemicals and Equipment

1. 6 M NH_3, ammonia water
2. Triethylamine
3. Aniline
4. N,N-Dimethylaniline
5. Acetamide
6. Diethyl ether (ether)
7. 6 M NaOH
8. Concentrated HCl
9. 6 M HCl
10. 6 M H_2SO_4
11. pH papers
12. Hot plate

NAME SECTION DATE

PARTNER GRADE

Experiment 7

PRE-LAB QUESTIONS

1. Draw the structure of the functional group that is found in an amine and an amide.

2. What is the general rule for the solubility of amines in water?

3. What happens when an amine is mixed with hydrochloric acid?

4. Compare the basicity of the following amines to ammonia: ethylamine (an aliphatic amine), aniline (an aromatic amine).

NAME ______ SECTION ______ DATE ______

PARTNER ______ GRADE ______

Experiment 7

REPORT SHEET

Properties of amines

	Odor		Solubility			pH
	Original Sol.	**with HCl**	**H_2O**	**Ether**	**HCl**	**H_2O**
6M NH_3						
Triethylamine						
Aniline						
N,N-Dimethylaniline						
Acetamide						

Triethylamine and concentrated hydrochloric acid observation:

Hydrolysis of acetamide

1. Acid solution
   a. pH reading:

   b. Odor noted:

2. Base solution
   a. pH reading:

   b. Odor noted:

POST-LAB QUESTIONS

1. The active ingredient in the commercial insect repellent "OFF" is N,N-diethyl-m-toluamide. Is this a primary, secondary or tertiary amide?

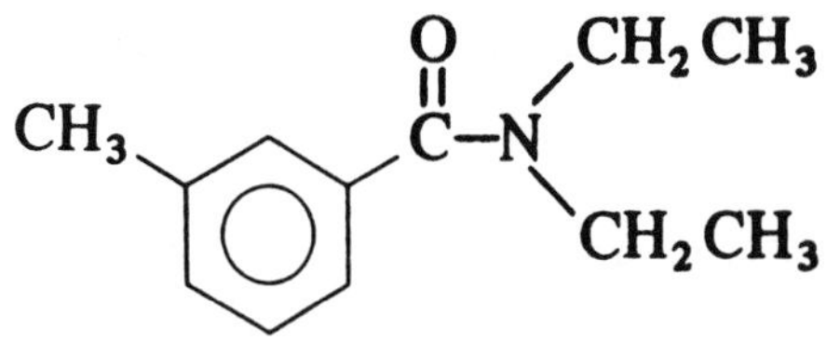

2. Of the amines tested which was the least soluble in water? Why?

3. Why does triethylamine lose its odor when mixed with hydrochloric acid?

4. Write the equations that accounts for what happens in the hydrolysis of the acetamide solution in a) acid and in b) base.

a.

b.

EXPERIMENT 8
Polymerization Reactions

Background

Polymers are giant molecules made of many (poly-) small units. The starting material, which is a single unit, is called the monomer. Many of the most important biological compounds are polymers. Cellulose and starch are polymers of glucose units, proteins are made of amino acids, and nucleic acids are polymers of nucleotides. Since the 1930s a large number of man-made polymers have been manufactured. They contribute to our comfort and gave rise to the previous slogan of DuPont Co.: "Better living through chemistry." Man-made fibers such as nylon and polyesters, plastics such as the packaging materials made of polyethylene and polypropylene films, polystyrene and polyvinyl chloride, just to name a few, all became household words. Man-made polymers are parts of buildings, automobiles, machinery, toys, appliances etc.; we encounter them daily in our life.

We focus our attention in this experiment on man-made polymers and the basic mechanism by which some of them are formed. The two most important types of reactions that are employed in polymer manufacturing are the addition and condensation polymerization reactions. The first is represented by the polymerization of styrene and the second by the formation of nylon.

Styrene is a simple organic monomer which by its virtue of containing a double bond can undergo addition polymerization

$$H_2C=CH + H_2C=CH \rightarrow H_3C-CH-CH=CH$$

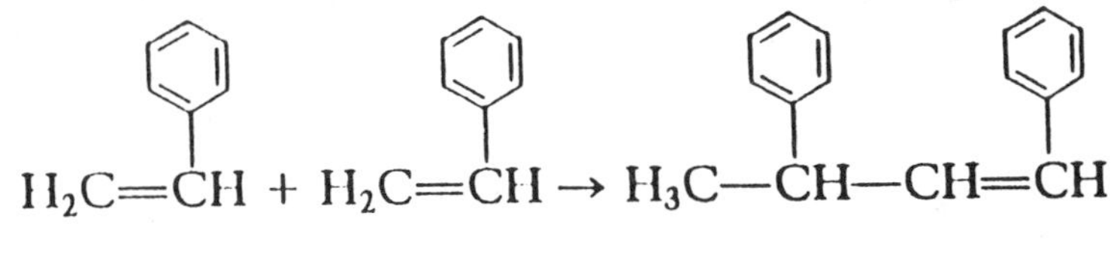

The reaction is called an addition reaction because two monomers are added to each other with the elimination of a double bond. However, the reaction as such does not go without the help of an unstable molecule, called an initiator, that starts the reaction. Benzoyl peroxide or t-butyl benzoyl peroxide are such initiators. Benzoyl peroxide splits into two halves under the influence of heat or U.V. light and thus produces two free radicals. A free radical is a molecular fragment that has one unpaired electron.

Thus, when the central bond was broken in the benzoyl peroxide each of the shared pair of electrons went with one half of the molecule, each containing an unpaired electron.

$$
\text{C}_6\text{H}_5-\overset{\overset{\text{O}}{\|}}{\text{C}}-\text{O}-\text{O}-\overset{\overset{\text{O}}{\|}}{\text{C}}-\text{C}_6\text{H}_5 \xrightarrow{\Delta} 2\ \text{C}_6\text{H}_5-\overset{\overset{\text{O}}{\|}}{\text{C}}-\text{O}^{\bullet}
$$

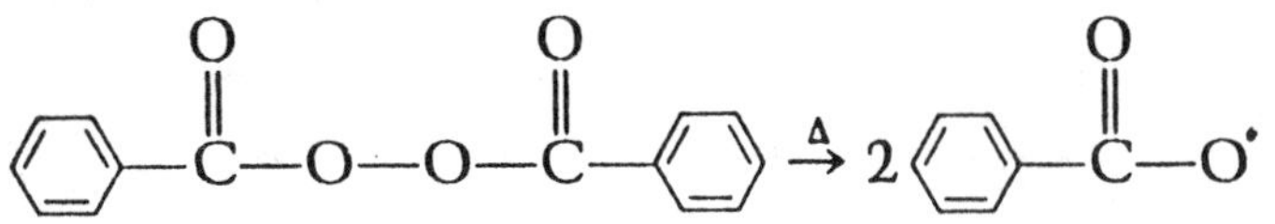

benzoyl peroxide

Similarly, tertiary butyl benzoyl peroxide also gives two free radicals:

$$
\text{C}_6\text{H}_5-\overset{\overset{\text{O}}{\|}}{\text{C}}-\text{O}-\text{O}-\overset{\overset{\text{CH}_3}{|}}{\underset{\underset{\text{CH}_3}{|}}{\text{C}}}-\text{CH}_3 \xrightarrow{\Delta} \text{C}_6\text{H}_5-\overset{\overset{\text{O}}{\|}}{\text{C}}-\text{O}^{\bullet} + \text{O}-\overset{\overset{\text{CH}_3}{|}}{\underset{\underset{\text{CH}_3}{|}}{\text{C}}}-\text{CH}_3
$$

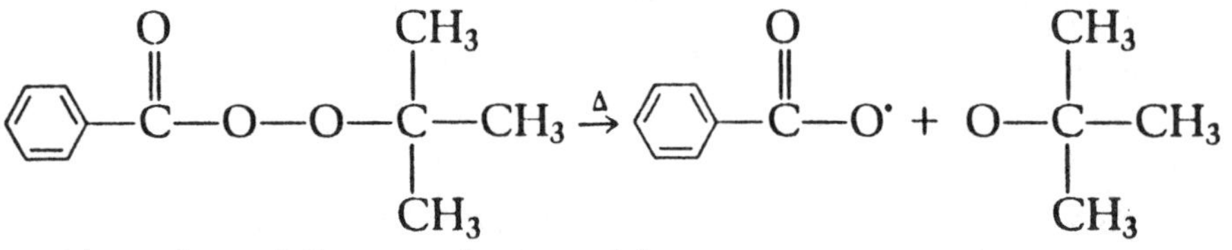

tertiary butyl benzoyl peroxide

The dot indicates the unpaired electrons. The free radical reacts with styrene and initiates the reaction:

$$
\text{C}_6\text{H}_5-\overset{\overset{\text{O}}{\|}}{\text{C}}-\text{O}^{\bullet} + \text{H}_2\text{C}{=}\overset{\overset{\text{C}_6\text{H}_5}{|}}{\text{CH}} \rightarrow \text{C}_6\text{H}_5-\overset{\overset{\text{O}}{\|}}{\text{C}}-\text{O}-\text{CH}_2-\overset{\overset{\text{C}_6\text{H}_5}{|}}{\text{CH}}^{\bullet}
$$

styrene

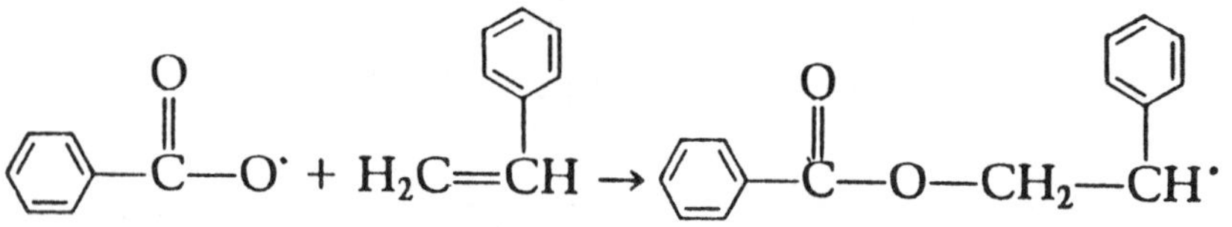

After this, the styrene monomers are added to the growing chain one by one until giant molecules containing hundreds and thousands of styrene-repeating units are formed. Please note the distinction between the monomer and the repeating unit. The monomer is the starting material, the repeating unit is part of the polymer chain. Chemically they are not identical. In the case of styrene the monomer contains a double bond, the repeating unit (in the brackets in the following structure) does not.

$$
\text{CH}_3-\overset{\overset{\text{C}_6\text{H}_5}{|}}{\text{CH}}-\left[\text{CH}_2-\overset{\overset{\text{C}_6\text{H}_5}{|}}{\text{CH}}\right]_n-\text{CH}_2-\overset{\overset{\text{C}_6\text{H}_5}{|}}{\text{CH}_2}
$$

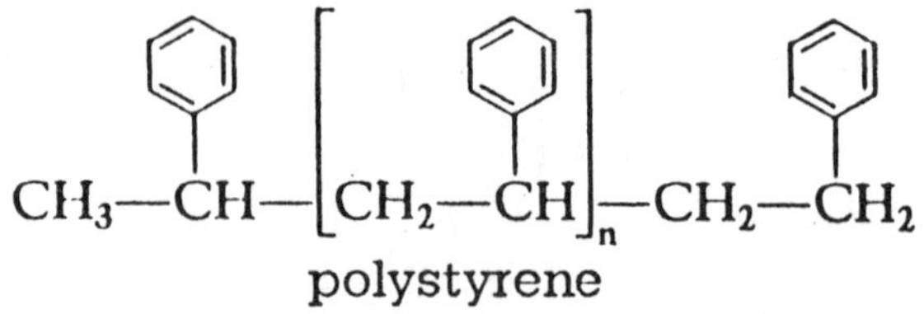

polystyrene

Since the initiators are unstable compounds, care should be taken not to keep them near flames or heat them directly. Even dropping the bottle containing a peroxide initiator may create a minor explosion.

The second type of reaction is called a condensation reaction because we condense two monomers into a longer unit and at the same time we eliminate- expel- a small molecule. Nylon 6-6 is made of adipoyl chloride and hexamethylene diamine:

$$n \; Cl-\overset{\overset{\displaystyle O}{\|}}{C}-CH_2-CH_2-CH_2-CH_2-\overset{\overset{\displaystyle O}{\|}}{C}-Cl + n \; H_2N-CH_2-CH_2-CH_2-CH_2-CH_2-CH_2-NH_2 \rightarrow$$

adipoyl chloride hexamethylene diamine

$$Cl-\overset{\overset{\displaystyle O}{\|}}{C}(-CH_2)_4-\overset{\overset{\displaystyle O}{\|}}{C}-\left[NH-(CH_2)_6-NH-\overset{\overset{\displaystyle O}{\|}}{C}-(CH_2)_4-\overset{\overset{\displaystyle O}{\|}}{C}- \right]_n NH-(CH_2)_6-NH_2 + n \; HCl$$

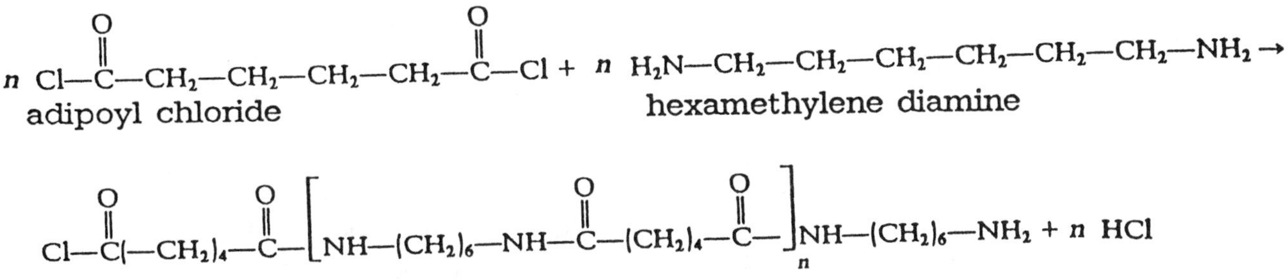

repeating unit

nylon 6-6

We form an amide linkage between the adipoyl chloride and the amine with the elimination of HCl. The polymer is called nylon 6-6 because there are six carbon atoms in the acyl chloride and six carbon atoms in the diamine. Other nylons, such as nylon 6-10, are made of sebacoyl chloride (a ten carbon containing acyl chloride) and hexamethylene diamine (six carbon atoms). We use acyl chloride rather than carboxylic acid to form the amide bond because the former is more reactive. NaOH is added to the polymerization reaction in order to neutralize the HCl that is released every time an amide bond is formed.

The length of the polymer chain formed in both reactions depends on environmental conditions. Usually the chains formed can be made longer by heating the products longer. This process is called curing.

Objectives

1. To acquaint students with the conceptual and physical distinction between monomer and polymer.
2. To perform addition and condensation polymerization and solvent casting of films.

Procedure

Preparation of Polystyrene

1. Set up your hot plate in the hood. Place 25 mL styrene in a 150-mL beaker. Add 20 drops of tertiary butyl benzoyl peroxide (tertiary butyl peroxide benzoate) initiator. Mix the solution.

2. Heat the mixture on the hot plate to about 140°C. The mixture will turn yellow.

3. When bubbles appear remove the beaker from the hot plate with beaker tongs. The polymerization reaction is exothermic and thus it generates its own heat. Overheating would create sudden boiling. When the bubbles disappear put the beaker back on the hot plate. But every time the mixture starts boiling you must remove the beaker.

4. Continue the heating until the mixture in the beaker has a syrupy consistency.

5. Pour the contents of the beaker onto a clean watch glass and let it solidify. The beaker can be cleaned from residual polystyrene by adding xylene and warming it on the hot plate under the hood until the polymer is dissolved.

6. Pour a few drops of the warm xylene solution on a microscopic slide and let the solvent evaporate. A thin film of polystyrene will be obtained. This is one of the techniques--the so-called solvent-casting technique--used to make films from bulk polymers.

7. Discard the remaining xylene solution into a special jar labeled "Waste." You can wash your beaker with soap and water.

8. Investigate the consistency of the solidified polystyrene on your watch glass. You can remove the solid mass by prying it off with a spatula.

Preparation of Nylon

1. Set up a 50-mL reaction beaker and clamp above it a cylindrical paper roll (from toilet paper) or a stick.

2. Add 2.0 mL of 20% NaOH solution and 10 mL of a 5% aqueous solution of hexamethylene diamine.

3. Take 10 mL of 5% adipoyl chloride solution in cyclohexane with a pipet or syringe. Layer the cyclohexane solution slowly on top of the aqueous solution in the beaker. Two layers will form and nylon will be produced at the interface (Fig. 8.1).

4. With a bent wire first scrape off the nylon formed on the walls of the beaker.

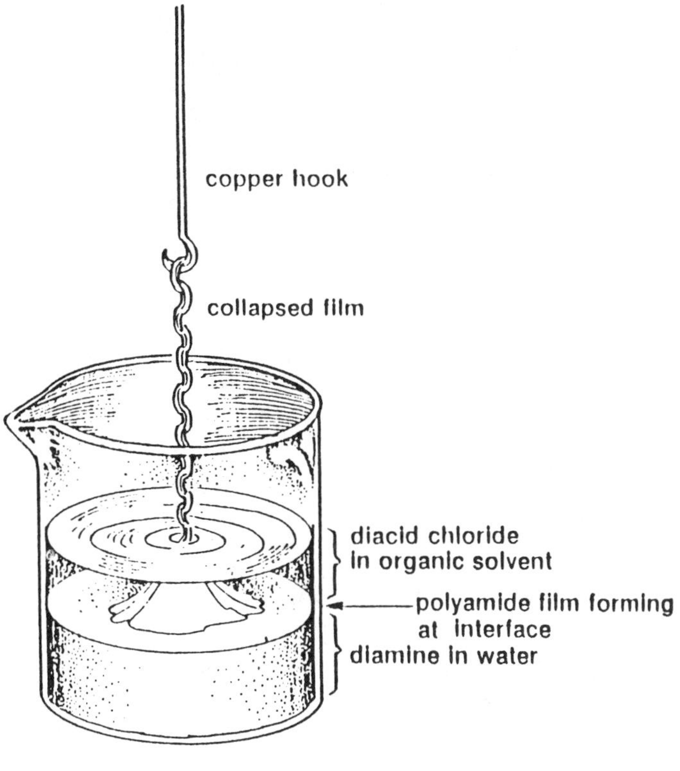

Figure 8.1 Preparation of nylon.

5. Then slowly lift the film from the center. Pull it slowly. If you pull it too fast, the nylon rope will break.

6. Wind it around the paper roll or stick two to three times. Do not touch it with your hands.

7. Slowly rotate the roll or the stick and wind at least a 1-meter nylon rope.

8. Cut the rope and transfer the wound rope into a beaker filled with water 4 (or 50% ethanol). Watch as the thickness of the rope collapses. Dry the rope between two filter papers.

9. There are still monomers left in the beaker. Mix the contents vigorously with a glass rod. Note the beads of nylon that have formed.

10. Pour the mixture into a cold water bath and wash it. Dry the nylon between two filter papers. Note the consistency of your products.

11. Dissolve a small amount of nylon in 80% formic acid. Place a few drops of the solution onto a microscope slide and evaporate the solvent under the hood.

12. Compare the appearance of the solvent cast nylon film with that of the polystyrene.

Chemicals and Equipment

1. Styrene
2. Hexamethylene diamine solution
3. Adipoyl chloride solution
4. Sodium hydroxide solution
5. Xylene
6. Formic acid solution
7. t-Butyl peroxide benzoate initiator
8. Hot plate
9. Paper roll or stick
10. Bent wires
11. 10 mL pipets
12. Spectroline pipet filler
13. Beaker tongs

NAME ______________________ SECTION ________ DATE ________

PARTNER ______________________ GRADE ______________________

Experiment 8

PRE-LAB QUESTIONS

1. Why should you not expose t-butyl peroxide to direct heat?

2. Write the structure of t-butyl peroxide.

3. Write the reaction for the polymerization of tetrafluoroethene. Show the repeating unit of the resulting polymer (Teflon).

4. Write the structure of the monomers and that of the repeating unit in nylon 6-10.

5. Why do we call nylon a condensation polymer?

NAME _____ SECTION _____ DATE _____

PARTNER _____ GRADE _____

Experiment 8

REPORT SHEET

1. Describe the appearance of polystyrene and nylon.

2. Could you distinguish between polystyrene and nylon on the basis of solubility?

3. Is there any difference in the appearance of the solvent cast films of nylon and polystyrene?

POST-LAB QUESTIONS

1. A polyester is made of sebacoyl chloride and ethylene glycol,

$$\underset{ClC(CH_2)_8CCl}{\overset{O\quad\quad O}{\parallel\quad\quad\parallel}} \quad + \quad \underset{CH_2OH}{\overset{CH_2OH}{|}} \quad \rightarrow$$

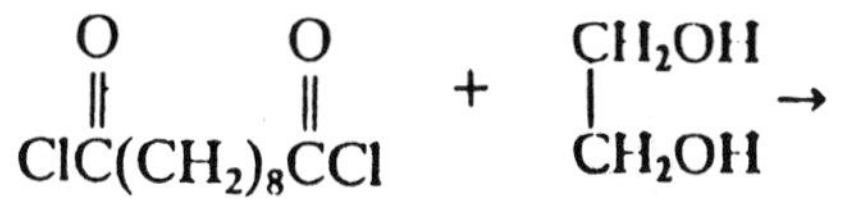

a. Draw the structure of the polyester formed.

Experiment 8

b. What molecules have been eliminated in this condensation reaction?

2. Two compounds , 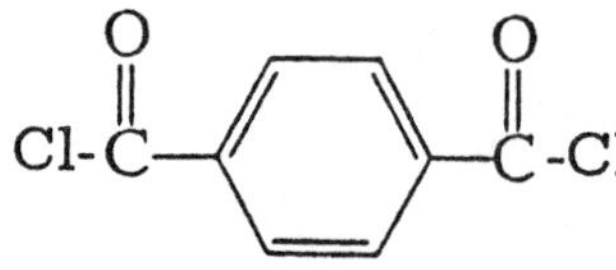

in cyclohexane and

$H_2N\text{-}(CH_2)_4\text{-}NH_2$ in water are reacted. Write the structure of the polyamide rope formed.

3. Since the polymerization of styrene is an exothermic reaction, why do you need to heat the mixture to 140°C?

EXPERIMENT 9
Preparation of aspirin

Background

One of the most widely used over-the-counter non-prescription drugs is aspirin. The United States alone sells more than 15,000 pounds each year. It is no wonder it is in such wide use when one considers how aspirin exerts its medicinal effect. It is an effective analgesic (pain killer) that can reduce the mild pain of headaches, toothache, neuralgia (nerve pain), muscle pain and joint pain (from arthritis and rheumatism). Aspirin behaves as an antipyretic drug (it reduces fever), and an anti-inflammatory agent capable of reducing the swelling and redness associated with inflammation.

Early studies showed the active agent that gave these properties was salicylic acid. However, salicylic acid contains the phenolic group and the carboxylic acid. As a result, the compound was too harsh to the linings of the mouth, esophagus and stomach. Contact with the stomach lining caused some hemorrhaging. The Bayer Company in Germany patented the ester acetylsalicylic acid and marketed the product as "aspirin" in 1899. Their studies showed that this material was less of an irritant; the acetylsalicylic acid was found to hydrolyze in the small intestine to salicylic acid and then was absorbed into the bloodstream. The relationship between salicylic acid and aspirin are shown in the following formulas:

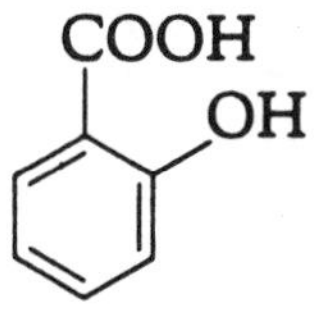

Salicylic acid

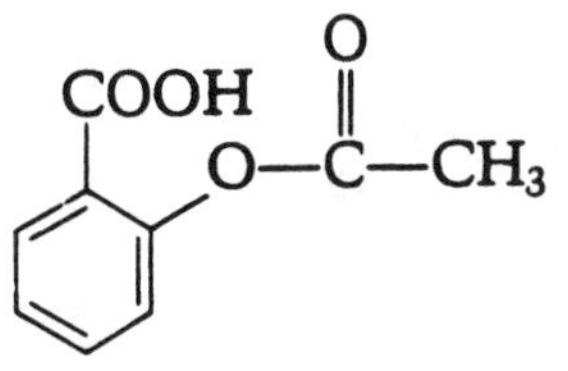

Aspirin (acetylsalicylic acid)

Aspirin still has its side effects. Note that the carboxylic acid functional group remains intact. As a result, hemorrhaging of the stomach walls can occur even with normal dosages. These side effects can be reduced through the use of buffering agents. Antacids, in the form of magnesium hydroxide, magnesium carbonate, and aluminum glycinate, when mixed into the formulation of the aspirin (e.g. Bufferin), reduce the acidic irritation.

This experiment will acquaint you with a simple synthetic problem in the preparation of aspirin. While the ester can be formed from acetic acid and salicylic acid, a better preparative method uses acetic anhydride in the reaction instead of acetic acid. An acid catalyst, like sulfuric or phosphoric acid, is used to speed up the process.

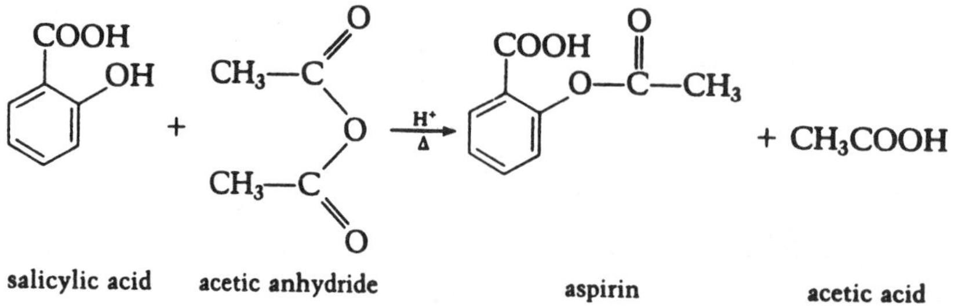

If any salicylic acid remains unreacted, its presence can be detected with a 1% ferric chloride solution. Salicylic acid has a phenol group in the molecule. The ferric chloride gives a violet color with any molecule possessing a phenol group (see Experiment 4). Notice the aspirin no longer has the phenol group. Thus, a pure sample of aspirin will not give a purple color with 1% ferric chloride.

Objectives

1. To illustrate the synthesis of the drug, aspirin.
2. To use a chemical test to determine the purity of the preparation.

Procedure

Preparation of aspirin

1. Prepare a bath using a 400-mL beaker filled about half-way with water. Heat to boiling.

2. Weigh 4.0 g of salicylic acid and place it in a 125-mL Erlenmeyer flask. Use this quantity of salicylic acid to calculate the theoretical or expected yield of aspirin (1). Carefully add 6 mL of acetic anhydride to the flask, and then while swirling, add 5 drops of concentrated sulfuric acid.

CAUTION! Acetic anhydride will irritate your eyes. Sulfuric acid will cause burns to the skin. Handle both chemicals with care. Dispense in the hood.

3. Mix the reagents and then place the flask in the boiling water bath; heat for 20 min. (Fig. 9.1). The solid will completely dissolve. Swirl the solution occasionally.

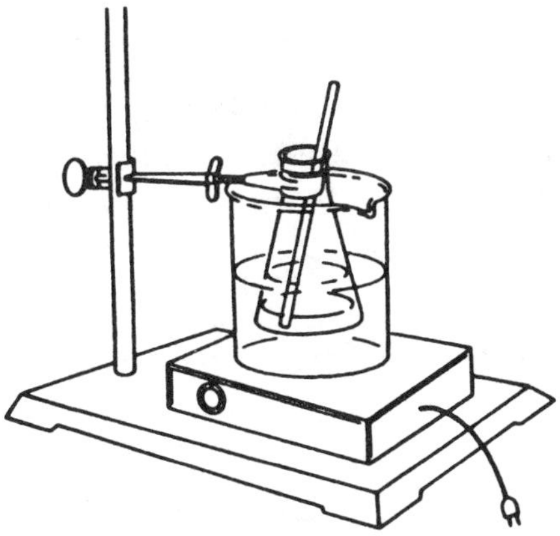

Figure 9.1 Assembly for the synthesis of aspirin.

4. Remove the Erlenmeyer flask from the bath and let it cool to approximately room temperature. Then slowly pour the solution into a 250-mL beaker containing 40 mL of ice water, mix thoroughly, and place the beaker in an ice bath. The water destroys any unreacted acetic anhydride and will cause the insoluble aspirin to precipitate from solution.

5. Collect the crystals by suction filtration with a Büchner funnel. The assembly is shown in Fig. 9.2 and is described in steps 6, 7 and 8 below.

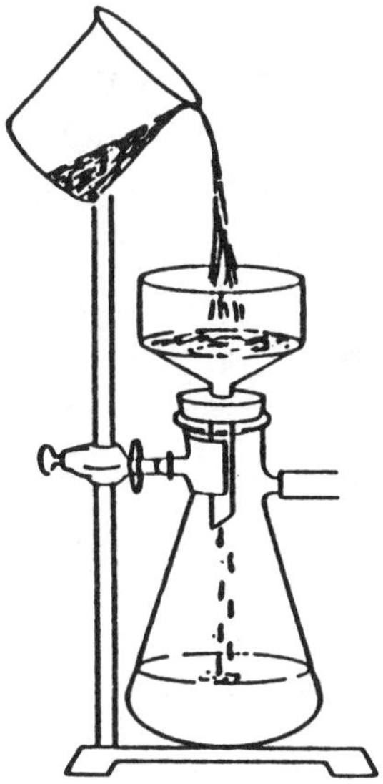

Figure 9.2 Filtering using the Büchner funnel.

6. Obtain a 500-mL filter flask and connect the side-arm of the filter flask to a water aspirator with heavy wall vacuum rubber tubing. (The thick walls of the tubing will not collapse when the water is turned on and the pressure is reduced.)

7. The Büchner funnel is inserted into the filter flask through either a filtervac, a neoprene adapter or a one-hole rubber stopper, whichever is available. Filter paper is then placed into the Büchner funnel. Be sure that the paper lies flat and covers all the holes. Wet the filter paper with water.

8. Turn on the water aspirator to maximum water flow. Pour the solution into the Büchner funnel.

9. Wash the crystals with two 10 mL portions of cold water followed by one 20 mL portion of cold ethanol.

10. Continue suction through the crystals for several minutes to help dry them. Disconnect the rubber tubing from the filter flask before turning off the water aspirator.

11. Using a spatula place the crystals between several sheets of paper toweling or filter paper and press-dry the solid.

12. Weigh a 50-mL beaker (2) to 0.001 g. Add the crystals and reweigh (3). Calculate the weight of crude aspirin (4). Determine the percent yield (5).

Determine the purity of the aspirin

1. The aspirin you prepared is not pure enough for use as a drug and is not suitable for ingestion. The purity of the sample will be tested with 1% ferric chloride solution and compared with a commercial aspirin and salicylic acid.

2. Label three test tubes (100 x 13 mm) no. 1, no. 2 and no. 3; place a few crystals of salicylic acid into test tube no. 1, a small sample of your aspirin into test tube no. 2, and a small sample of a crushed commercial aspirin into test tube no. 3. Add 5 mL of distilled water to each test tube and swirl to dissolve the crystals.

3. Add 10 drops of 1% aqueous ferric chloride to each test tube.

4. Compare and record your observations. The formation of a purple color indicates the presence of salicylic acid. The intensity of the color qualitatively tells how much salicylic acid is present.

Chemicals and Equipment

1. Bunsen burner
2. Boiling chips
3. Büchner funnel, small
4. 500-mL Filter flask
5. Filter paper
6. Filtervac or neoprene adaptor
7. Salicylic acid
8. Commercial aspirin tablets
9. 1% ferric chloride solution
10. Acetic anhydride
11. Concentrated sulfuric acid
12. 95% ethanol

NAME ______________________ SECTION __________ DATE __________

PARTNER ______________________ GRADE ______________________

Experiment 9

PRE-LAB QUESTIONS

1. A person took aspirin for its antipyretic effect. What kind of malady might the person be trying to cure?

2. Draw the structure of aspirin. Should this compound test positive with 1% ferric chloride solution? Explain your answer.

3. Aspirin can irritate the stomach. What group is responsible for the harshness of the drug?

4. What is the active ingredient that gives aspirin its therapeutic properties?

NAME ______________________ SECTION __________ DATE __________

PARTNER ______________________ GRADE __________

Experiment 9

REPORT SHEET

1. Theoretical yield:

$$\_\_\_\_\_\text{ g salicylic acid} \times \frac{180\text{ g aspirin}}{1\text{ mole}} \times \frac{1\text{ mole}}{138\text{ g salicylic acid}}$$

$$= \_\_\_\_\_\_\_\_\_\_\text{ g aspirin}$$

2. Weight of 50-mL beaker __________ g

3. Weight of your aspirin and beaker __________ g

4. Weight of your aspirin: (3) - (2) __________ g

5. Percent yield: [(4)/(1)] x 100 __________ %

6. Ferric chloride test

No.	Sample	Color	Intensity
1	Salicylic acid		
2	Your aspirin		
3	Commercial aspirin		

POST-LAB QUESTIONS

1. What is the purpose of the concentrated sulfuric acid in the preparation of aspirin? Could some other acid be used?

2. If in step no. 11 of the procedure, you failed to dry completely your aspirin preparation by omitting the drying between filter paper, what would happen to your percent yield?

3. A student expected 8.4 g of acetylsalicylic acid, but obtained only 5.3 g. What is the percentage yield?

4. Tylenol also is an analgesic often taken by people allergic to aspirin. The active ingredient is acetaminophen.

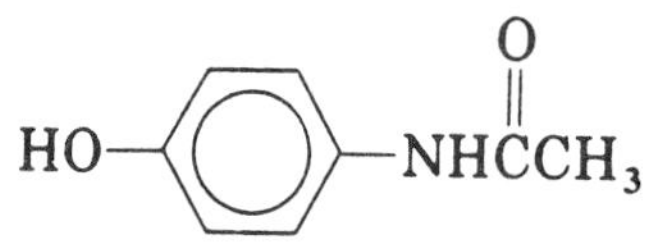

Acetaminophen

Would acetaminophen give a positive phenol test? Explain your answer. Does this compound contain the ester functional group? If not, what functional group is present?

EXPERIMENT 10

Isolation of caffeine from tea leaves

Background

Many organic compounds are obtained from natural sources through extraction. This method takes advantage of the solubility characteristics of a particular organic substance with a given solvent. In the experiment here, caffeine is readily soluble in hot water and is thus, separated from the tea leaves. Caffeine is one of the main substances that make up the water solution called tea. Besides being found in tea leaves, caffeine is present in coffee, kola nuts, and cocoa beans. As much as 5% by weight of the leaf material in tea plants consists of caffeine.

The caffeine structure is shown below. It is classed as an

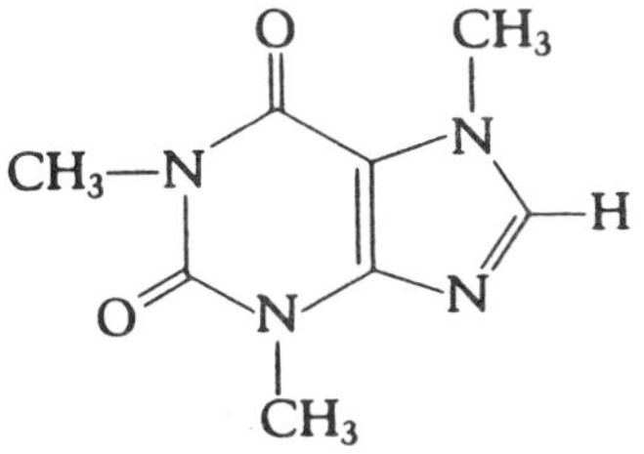

alkaloid, meaning that with the nitrogen present, the molecule has base characteristics (alkali-like). In addition the molecule has the purine ring system, a framework which plays an important role in living systems.

Caffeine is the most widely used of all the stimulants. Small doses of this chemical (50 to 200 mg) can increase alertness and reduce drowsiness and fatigue. The popular "No-Doz" tablet has caffeine as the main ingredient. In addition it affects blood circulation since the heart is stimulated and blood vessels are relaxed (vasodilation). It also acts as a diuretic. There are side effects. Large doses of over 200 mg can result in insomnia, restlessness, headaches, and muscle tremors ("coffee nerves"). Continued, heavy use may bring on physical dependence. (How many of you know someone who cannot function in the morning until they have that first cup of coffee?)

Tea leaves consist primarily of cellulose since this is the principle structural material of all plant cells. Fortunately, the cellulose is insoluble in water, so that by using a hot water extraction, the more soluble caffeine can be separated. Also dissolved in water are complex substances called tannins. These are colored phenolic compounds of high molecular weight (500 to 3000) that have acidic behavior. If a basic salt such as Na_2CO_3 is added to the water solution, the tannins can react to form a salt. These salts are insoluble in organic solvents such as chloroform or methylene chloride, but are soluble in water.

Although caffeine is soluble in water (2 g/100 g of cold water), it is more soluble in the organic solvent methylene chloride (14 g/100 g). Thus, caffeine can be extracted from the basic tea solution with methylene chloride, but the sodium salts of the tannins remain behind in the aqueous solution. Evaporation of the methylene chloride yields the caffeine.

Objectives

1. To demonstrate the isolation of a natural product.
2. To learn the techniques of extraction.

Procedure

The isolation of caffeine from tea leaves follows the scheme below:

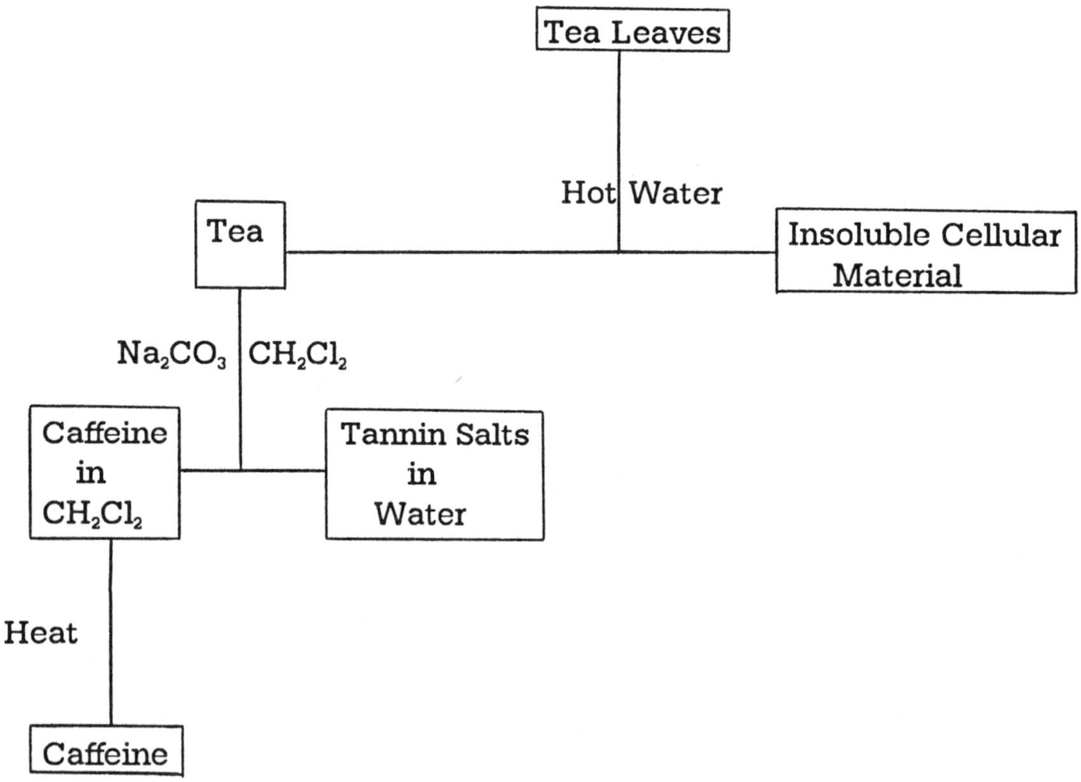

1. Carefully open two commercial tea bags (try not to tear the paper) and weigh the contents to the nearest 0.001 g. Record this weight (1). Place the tea leaves back into the bags, close and secure the bags with staples.

2. Into a 150-mL beaker place the tea bags so that they lie flat on the bottom. Add 25 mL of distilled water and 2.0 g of anhydrous Na_2CO_3; heat the contents with a hot plate, keeping a gentle boil, for 20 min. While the mixture is boiling, keep a watch glass on the beaker. Hold the tea bags under water by occasionally pushing them down with a glass rod.

3. Decant the hot liquid into a 50-mL Erlenmeyer flask. Wash the tea bags with 10 mL of hot water, carefully pressing the tea bag with a glass rod; add this wash water to the tea extract. (If any solids are present in the tea extract, filter them by gravity to remove.) Cool the combined tea extract to room temperature. The tea bags may be discarded.

4. Transfer the cool tea extract to a 125-mL separatory funnel supported on a ring stand with a ring clamp.

5. Carefully add 5.0 mL of methylene chloride to the separatory funnel. Stopper the funnel and lift the funnel from the ring clamp; hold the funnel with two hands as shown in Fig. 10.1. By holding the stopper in place with one hand, invert the funnel. Make certain the stopper is held tightly and no liquid is spilled; open the stop-cock, being sure to point the opening away from you and your neighbors. Built-up pressure caused by gases accumulating inside will be released. Now close the stop-cock and gently mix the contents by inverting two or three times. Again release any pressure by opening the stop-cock as before.

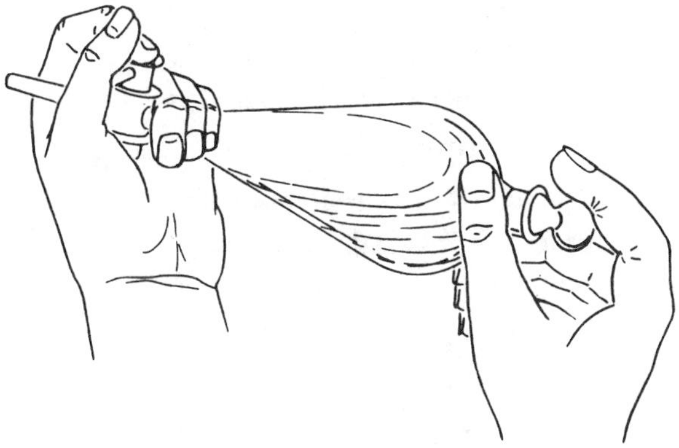

Figure 10.1 Using the separatory funnel.

6. Return the separatory funnel to the ring clamp, remove the stopper and allow the aqueous layer to separate from the methylene chloride layer (Fig. 10.2). You should see two distinct layers form after a few minutes with the methylene chloride layer at the bottom. Sometimes an emulsion may form at the juncture of the two layers. The emulsion often can be broken by gently swirling the contents or by gently stirring the emulsion with a glass rod.

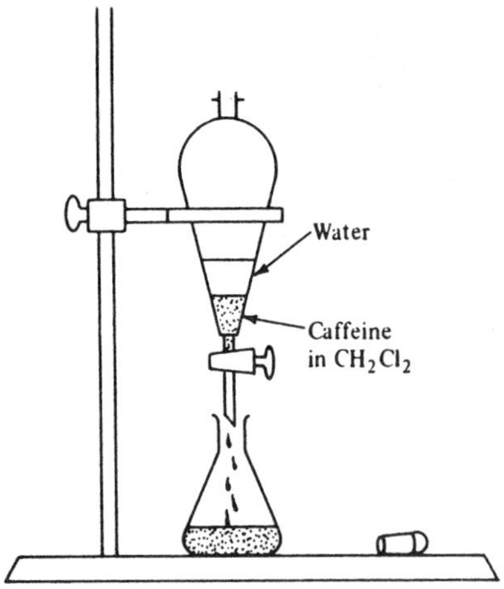

Figure 10.2 Separation of the aqueous layer and the methylene chloride layer in the separatory funnel.

7. Carefully drain the lower layer into a 25-mL Erlenmeyer flask. Try not to include any water with the methylene chloride layer; careful manipulation of the stop-cock will prevent this.

8. Repeat the extraction with an additional 5.0 mL of methylene chloride. Combine the separated bottom layer with the methylene chloride layer obtained from step 7.

9. Add 0.5 g of anhydrous Na_2SO_4 to the combined methylene chloride extracts. Swirl the flask. The anhydrous salt is a drying agent and will remove any water that may still be present.

10. Weigh a 25-mL Erlenmeyer flask containing one or two boiling stones. Record this weight (2). By means of a gravity filtration, filter the methylene chloride-salt mixture into the pre-weighed flask. Rinse the salt on the filter paper with an additional 2.0 mL of methylene chloride.

11. Remove the methylene chloride by evaporation <u>in the hood</u>. Be careful not to overheat the solvent since it may foam over. The solid residue which remains after the solvent is gone is the crude caffeine. Reweigh the cooled flask (3). Calculate the weight of the caffeine by subtraction (4) and determine the percent yield (5).

12. Take a melting point of your solid. First scrape the caffeine from the sides of the flask with a microspatula and collect the solid in a capillary tube. Pure caffeine melts at 238°C. Compare your melting point (6) to the literature value.

13. Collect the caffeine in a sample vial and submit it to your instructor.

Chemicals and Equipment

1. Hot plate
2. 125-mL separatory funnel with stopper
3. Melting point capillaries
4. Small sample vials
5. Filter paper, (7.0 cm) fast flow
6. Tea bags
7. Boiling chips
8. Anhydrous sodium sulfate, Na_2SO_4
9. Anhydrous sodium carbonate, Na_2CO_3
10. Methylene chloride, CH_2Cl_2
11. Stapler

NAME SECTION DATE

PARTNER GRADE

Experiment 10

PRE-LAB QUESTIONS

1. Name some sources of caffeine.

2. What part of the caffeine structure gives the alkaloid properties?

3. How does caffeine affect an individual?

4. How can you test the purity of a sample of caffeine?

NAME ____________________ SECTION __________ DATE __________

PARTNER ____________________ GRADE __________

Experiment 10

REPORT SHEET

1. Weight of tea in 2 tea bags ________g
2. Weight of 25-mL Erlenmeyer flask and boiling stones ________g
3. Weight of flask, boiling stones and crude caffeine ________g
4. Weight of caffeine: (3) - (2) ________g
5. Percent yield: [(4)/(1)] x 100 ________%
6. Melting point of your caffeine ________°C

POST-LAB QUESTIONS

1. Explain why the melting point of your caffeine sample was lower than the literature value?

2. What other compounds were extracted along with caffeine in the hot water extract (tea)? How were these compounds separated from the caffeine?

3. In the procedure, why was the caffeine eventually recovered from the methylene chloride extract and not directly from the water?

4. How do you know that methylene chloride has a greater density than water?

EXPERIMENT 11
Carbohydrates

Background

Carbohydrates are polyhydroxy aldehydes or ketones or compounds that yield polyhydroxy aldehydes or ketones upon hydrolysis. Rice, potatoes, bread, corn, candy, and fruits are rich in carbohydrates. A carbohydrate can be classified as a monosaccharide (for example glucose or fructose), a disaccharide (sucrose or lactose), which consists of two joined monosaccharides, or a polysaccharide (starch or cellulose), which consists of thousands of monosaccharide units linked together. Monosaccharides exist mostly as cyclic structures containing hemiacetal or hemiketal groups. These structures in solutions are in equilibrium with the corresponding open chain structures bearing aldehyde or ketone groups. Glucose, blood sugar, is an example of a polyhydroxy aldehyde (Fig. 11.1).

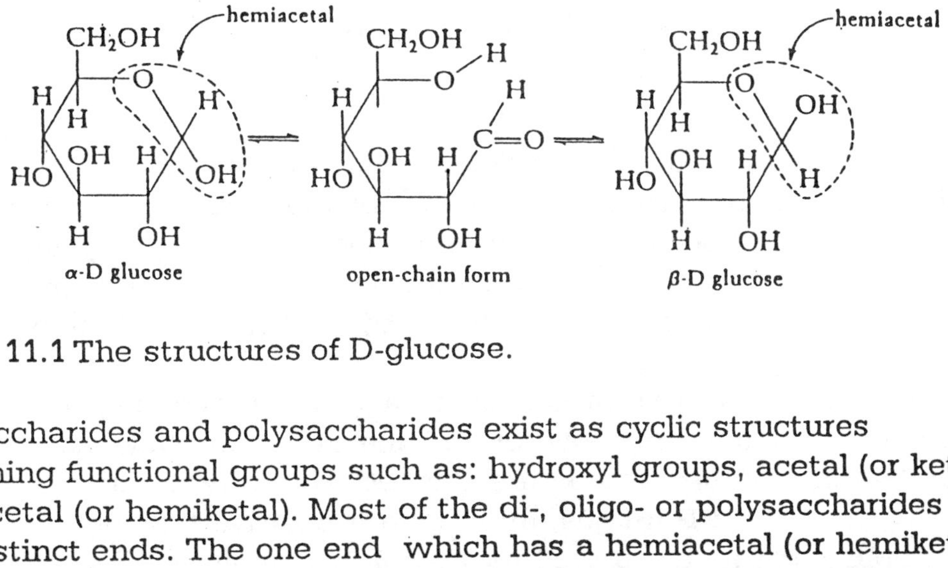

Figure 11.1 The structures of D-glucose.

Disaccharides and polysaccharides exist as cyclic structures containing functional groups such as: hydroxyl groups, acetal (or ketal), hemiacetal (or hemiketal). Most of the di-, oligo- or polysaccharides have two distinct ends. The one end which has a hemiacetal (or hemiketal) on its terminal is called the reducing end, and the one which does not contain a hemiacetal (or hemiketal) terminal is the non-reducing end. The name "reducing" is given because hemiacetals and to a lesser extent hemiketals can reduce an oxidizing agent such as Benedict's reagent.

The following is an example:

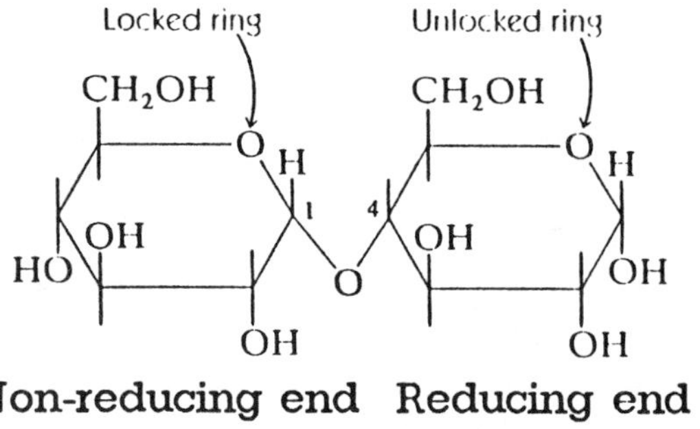

Figure 11.2 The structure of maltose, a disaccharide.

Not all disaccharides or polysaccharides contain a reducing end. An example is sucrose which does not have a hemiacetal or hemiketal group on either of its ends.

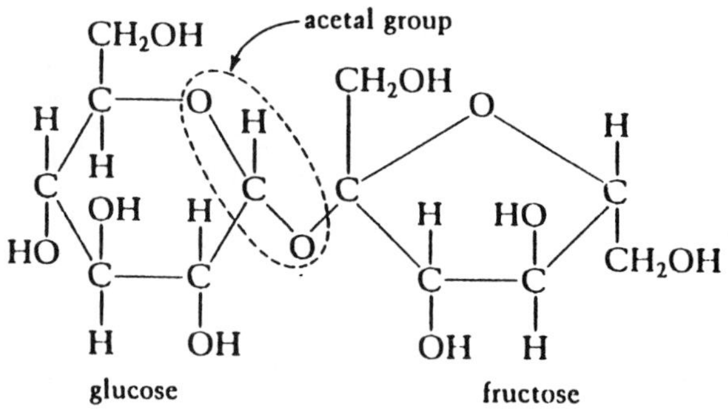

Figure 11.3 The structure of sucrose.

Polysaccharides, such as amylose or amylopectin do have a hemiacetal group on one of their terminal ends, but practically they are non-reducing substances, because there is only one reducing group for every 2-10,000 monosaccharidic units. In such a low concentration the reducing group does not give a positive test with Benedict's or Fehling's reagent.

On the other hand, when a non-reducing disaccharide (sucrose) or a polysaccharide such as amylose is hydrolyzed the glycosidic linkages (acetal) are broken and reducing ends are created. Hydrolyzed sucrose (a mixture of D-glucose and D-fructose) will give a positive test with Benedict's or Fehling's reagent as well as hydrolyzed amylose (a mixture of glucose and glucose containing oligosaccharides).

The hydrolysis of sucrose or amylose can be achieved by using a strong acid such as HCl or with the aid of biological catalysts (enzymes).

Starch can form an intense, brilliant, dark blue or violet colored complex with iodine. The straight chain component of starch, the amylose, gives a blue color while the branched component, the amylopectin yields a purple color. In the presence of iodine the amylose forms helixes inside of which the iodine molecule assemble as long polyiodide chains. The helix forming branches of amylopectin are much shorter than those of amylose. Therefore the polyiodide chains are also much shorter in the amylopectin-iodine complex than in the amylose-iodine complex. The result is a different color (purple). When starch is hydrolyzed and broken down to small carbohydrate units the iodine will not give a dark blue (or puple) color. The iodine test is used in this experiment to indicate the completion of the hydrolysis.

In this experiment you will investigate some chemical properties of carbohydrates in terms of their functional groups.

1. <u>Reducing and non-reducing properties of carbohydrates</u>

   a. **Aldoses (polyhydroxy aldehydes).** All aldoses are reducing sugars because they contain free aldehyde functional groups. The aldehydes are oxidized by mild oxidizing agents (e.g., Benedict's or Fehling's reagent) to the corresponding carboxylates. For example,

$$R\text{—}CHO + 2Cu^{2+} \xrightarrow{NaOH} R\text{—}COO^{-}Na^{+} + Cu_2O\downarrow$$

(from Fehling's reagent) red precipitate

   b. **Ketoses (polyhydroxy ketones).** All ketoses are reducing sugars because they have a ketone functional group next to an alcohol functional group. The reactivity of this specific ketone (also called -hydroxyketone) is attributed to its ability to form an -hydroxyaldehyde in basic media according to the following equilibrium equations:

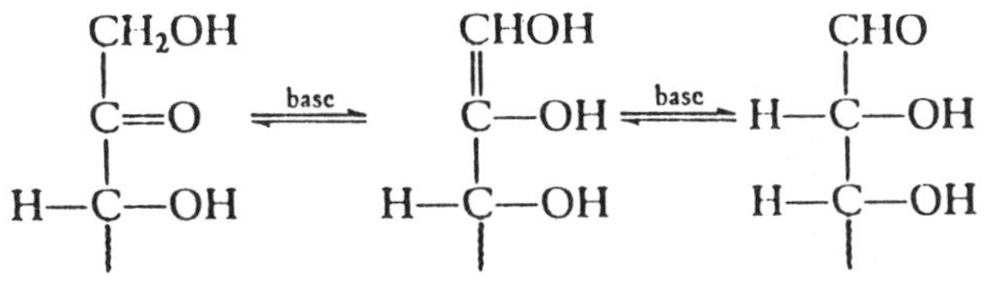

ketose enediol aldose

   c. **Hemiacetal functional group (potential aldehydes).** Carbohydrates with hemiacetal functional groups can reduce mild oxidizing agents such as Benedict's reagent because hemiacetals can easily form aldehydes through the following equiliblium equation:

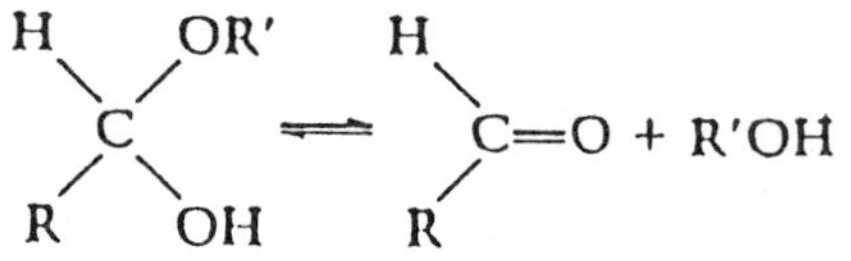

Sucrose is, on the other hand, a nonreducing sugar because it does not contain a hemiacetal functional group. Although starch has a hemiacetal functional group at one end of its molecule, it is, however, considered as a nonreducing sugar because the effect of the hemiacetal group in a very large starch molecule becomes insignificant to give a positive Benedict's test.

2. Hydrolysis of acetal groups. Disaccharides and polysaccharides can be converted into monosaccharides by hydrolysis. The following is an example:

$$C_{12}H_{22}O_{11} + H_2O \xrightarrow{\text{catalyst}} C_6H_{12}O_6 + C_6H_{12}O_6$$

lactose (milk sugar) — glucose — galactose

Objectives

1. To become familiar with the reducing or nonreducing nature of carbohydrates.
2. To experience the enzyme-catalyzed and acid-catalyzed hydrolysis of acetal groups.

Procedure

Reducing or nonreducing carbohydrates

Place approximately 2 mL (30 drops) of Fehling's solution (15 drops each of solution part A and solution part B) into each of five labeled tubes. Add ten drops of each of the following carbohydrates to the corresponding test tubes as shown in the following table.

Test tube no.	Name of carbohydrate
1	Glucose
2	Fructose
3	Sucrose
4	Lactose
5	Starch

Place the test tubes in a boiling water bath for 5 min. A 600-mL beaker containing about 200 mL of tap water and a few boiling chips is used as the bath. Record your results. Which of those carbohydrates are reducing carbohydrates?

Hydrolysis of carbohydrates

Hydrolysis of sucrose (Acid versus base catalysis)

Place 3 mL of 2% sucrose solution in each of two labeled test tubes. To the first test tube (no. 1) add 3 mL of water and three drops of dilute sulfuric acid solution. To the second test tube (no. 2) add 3 mL of water and three drops of dilute sodium hydroxide solution. Heat the test tubes in a boiling water bath for about 5 min. Cool both solutions to room temperature and then to the contents of test tube no. 1, add dilute sodium hydroxide solution (about ten drops) until red litmus paper turns blue. Test a few drops of each of the two solutions (test tubes nos. 1 and 2) with Fehling's reagent as described before. Record your results.

Hydrolysis of starch (Enzyme versus acid catalysis)

Place 2 mL of 2% starch solution in each of two labeled test tubes. To the first test tube (no. 1), add 2 mL of your own saliva. (Use a 10-mL graduated cylinder to collect your saliva.) To the second test tube (no. 2), add 2 mL of dilute sulfuric acid. Place both test tubes in a water bath that has been previously heated to 45°C. Allow the test tubes with their contents to stand in the warm water bath for 30 min. Transfer a few drops of each solution into a spot plate or two labeled microtest tubes. (Use two clean, separate medicine droppers for transferring.) To each test tube, add two drops of iodine solution. Record the color of the solutions.

Acid catalyzed hydrolysis of starch

Place 50 mL of starch solution in a 125-mL Erlenmeyer flask and add 10 mL of dilute sulfuric acid. Heat the solution in a boiling water bath for about 5 min. Using a clean medicine dropper, transfer about 3 drops of the starch solution into a spot plate or a microtest tube and then add 2 drops of iodine solution. Observe the color of the solution. If the solution gives a positive test with iodine solution (the solution should turn blue), continue heating. Transfer about 3 drops of the boiling solution at 5-min intervals for an iodine test. (Note: Rinse the medicine dropper very thoroughly before each test). When the solution no longer gives a blue color with iodine solution, stop heating and record the time needed for the completion of hydrolysis.

Chemicals and Equipment

1. Bunsen burner
2. Medicine droppers
3. Microtest tubes or white spot plates
4. Boiling chips
5. Fehling's reagent
6. 3 M NaOH
7. 2% starch solution
8. 2% sucrose
9. 2% fructose
10. 2% glucose
11. 2% lactose
12. 3 M H_2SO_4
13. 0.01 M iodine in KI

NAME ______________________ SECTION __________ DATE __________

PARTNER ______________________ GRADE __________

Experiment 11

PRE-LAB QUESTIONS

1. Circle the hemiacetal functional group and acetal functional group of the following carbohydrates:

a. sucrose

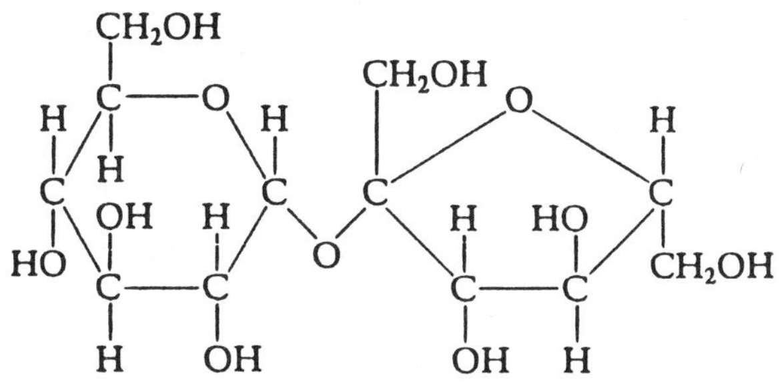

b. lactose

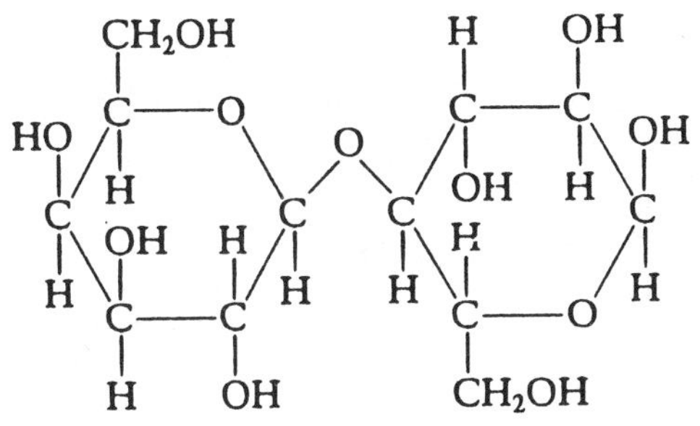

2. Which carbohydrate in question 1 is a reducing sugar?

3. Write an equation to indicate the formation of the products resulting from the dilute acid hydrolysis of each of the following carbohydrates: a. maltose ; b. sucrose

NAME ______ SECTION ______ DATE ______

PARTNER ______ GRADE ______

Experiment 11

REPORT SHEET

Reducing or nonreducing carbohydrates

Test tube no	Substance	Reducing or nonreducing carbohydrates
1	Glucose	
2	Fructose	
3	Sucrose	
4	Lactose	
5	Starch	

Hydrolysis of carbohydrates

Hydrolysis of sucrose (Acid versus base catalysis)

Sample	Condition of hydrolysis	Fehling's reagent (positive or negative)
1	Acidic (H_2SO_4)	
2	Basic (NaOH)	

Hydrolysis of starch (Enzyme versus acid catalysis)

Sample	Condition of hydrolysis	Iodine test (positive or negative)
1	Enzymatic (saliva)	
2	Acidic (H_2SO_4)	

Acid catalyzed hydrolysis of starch

Test tube no	Heating time (min)	Iodine test (positive or negative)
1	5	
2	10	
3	15	
4	20	

POST-LAB QUESTIONS

1. Fructose is a ketose. Ketones usually do not give a positive reaction with Fehling's reagent. Why is fructose oxidized by Fehling's reagent? (Consult your textbook).

2. What chemical test is performed to indicate the completion of the hydrolysis of starch by acid?

3. Which hydrolysis of starch is faster: the acid or the enzyme catalyzed reaction?

4. In Fig.16.13 of your textbook the structure of amylose is depicted. Amylose gives a positive iodine test and a negative test with Fehling's reagent. On the basis of the structure of amylose explain the results of the two tests.

EXPERIMENT 12
Preparation and properties of a soap

Background

A soap is the sodium or potassium salt of a long-chain fatty acid. The fatty acid usually contains 12 to 18 carbon atoms. Solid soaps usually consist of sodium salts of fatty acids, whereas liquid soaps consist of the potassium salts of fatty acids.

A soap, such as sodium stearate, consists of a nonpolar end (the hydrocarbon chain of the fatty acid) and a polar end (the ionic carboxylate).

$$CH_3CH_2CH_2CH_2CH_2CH_2CH_2CH_2CH_2CH_2CH_2CH_2CH_2CH_2CH_2CH_2CH_2 \overset{O}{\underset{\|}{-C}} -O^- Na^+$$

nonpolar	polar
(dissolves in oils)	(dissolves in water)

Because "like dissolves like," the nonpolar end (hydrophobic or water hating part) of the soap molecule can dissolve the greasy dirt, and the polar or ionic end (hydrophilic or water loving part) of the molecule is attracted to water molecules. Therefore, the dirt from the surface being cleaned will be pulled away and suspended in water. Thus, soap acts as an emulsifying agent, a substance used to disperse one liquid (oil molecules) in the form of finely suspended particles or droplets in another liquid (water molecules).

Treatment of fats or oils with strong bases such as lye (NaOH) or potash (KOH) causes them to undergo hydrolysis (saponification) to form glycerol and the salt of a long-chain fatty acid (soap).

$$\begin{array}{l} CH_2-O-\overset{O}{\overset{\|}{C}}-C_{17}H_{35} \\ \\ CH-O-\overset{O}{\overset{\|}{C}}-C_{17}H_{35} \\ \\ CH_2-O-\overset{O}{\overset{\|}{C}}-C_{17}H_{35} \end{array} + 3\,NaOH \xrightarrow{\Delta} \begin{array}{l} CH_2OH \\ | \\ CHOH \\ | \\ CH_2OH \end{array} + 3C_{17}H_{35}\overset{O}{\overset{\|}{C}}-O^-Na^+$$

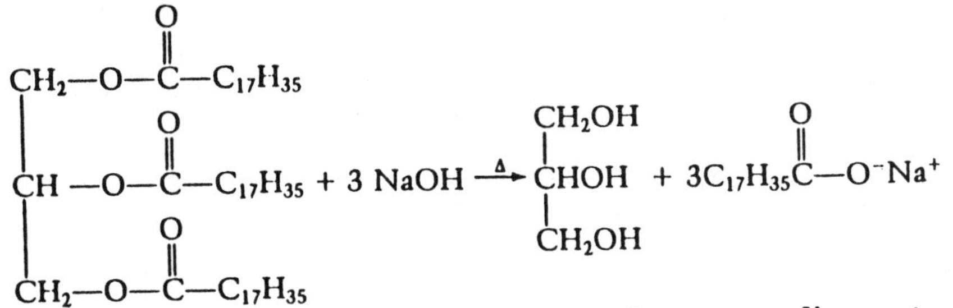

tristearin	glycerol	sodium stearate (a soap)

Because soaps are salts of strong bases and weak acids, they should be weakly alkaline in aqueous solution. However, a soap with free alkali can cause damage to skin, silk, or wool. Therefore, a test for basicity of the soap is quite important.

Soap has been largely replaced by synthetic detergents during the last two decades because soap has two serious drawbacks. One is that soap becomes ineffective in hard water, which contains appreciable amounts of Ca^{2+} or Mg^{2+}.

$$2C_{17}H_{35}COO^-Na^+ + M^{2+} \rightarrow (C_{17}H_{35}COO)_2^- M^{2+} \downarrow + 2Na^+$$

soap scum

M = (Ca^{2+} or Mg^{2+})

The other is that, in an acidic solution, soap is converted to free fatty acid and therefore loses its cleansing action.

$$C_{17}H_{35}COO^-Na^+ + H^+ \rightarrow C_{17}H_{35}COOH \downarrow + Na^+$$

soap fatty acid

Objectives

1. To prepare a simple soap.
2. To investigate some properties of a soap.

Procedure

Preparation of a soap

Measure 23 mL of a vegetable oil into a 250-mL Erlenmeyer flask. Add 20 mL of ethyl alcohol (to act as a solvent) and 20 mL of 25% sodium hydroxide solution. While stirring the mixture constantly with a glss rod, the flask with its contents is heated gently in a boiling water bath. A 600-mL beaker containing about 200 mL of tap water and a few boiling chips can serve as a water bath (Fig. 12.1).

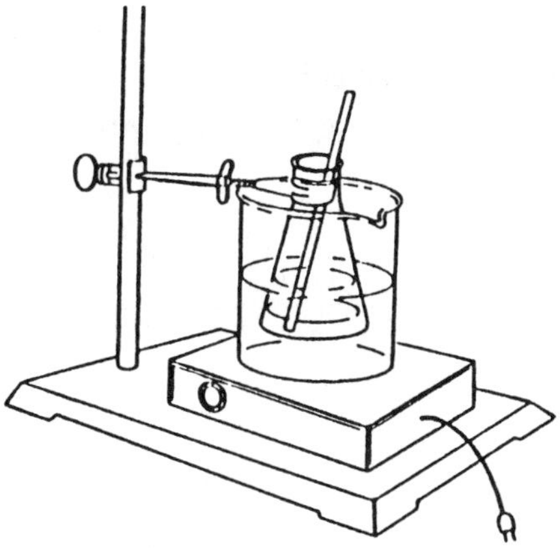

12.1 Experimental set-up for soap preparation.

CAUTION! Alcohol is flammable!

After being heated for about 20 min, the odor of alcohol will disappear, indicating the completion of the reaction. A pasty mass containing a mixture of the soap, glycerol, and excess sodium hydroxide is obtained. Use an ice-water bath to cool the flask with its contents. To precipitate or "salt out" the soap, add 150 mL of a saturated sodium chloride solution to the soap mixture while stirring vigorously. This process increases the density of the aqueous solution; therefore, soap will float out from the aqueous solution. Filter the precipitated soap with the aid of suction and wash it with 10 mL of ice cold water. Observe the appearance of your soap and record your observation.

Properties of a soap

1. Emulsifying Properties. Shake five drops of mineral oil in a test tube containing 5 mL of water. A temporary emulsion of tiny oil droplets in water will be formed. Repeat the same test, but this time, add a small piece of the soap you have prepared before shaking. Allow both solutions to stand a while. Compare the appearance and the relative stabilities of the two emulsions.Record your observations on the Report Sheet.

2. Hard Water Reactions. Place about one-third spatulaful of the soap you have prepared in a 50-mL beaker containing 25 mL of water. Warm the beaker with its contents to dissolve the soap. Pour 5 mL of a soap solution into each of five test tubes (nos. 1, 2, 3, 4, and 5). Test no. 1 with two drops of a 5% solution of calcium chloride, no. 2 with two drops of a 5% solution of magnesium chloride, no. 3 with two drops of a 5% solution of iron chloride, and no. 4 with tap water. The no. 5 tube will be used for a basicity test which will be performed later. Record your observations on the Report Sheet.

3. Alkalinity (Basicity). Test the soap solution no. 5 with a wide-range pH paper. What is the approximate pH of your soap solution? Record your answer on the Report Sheet.

Chemicals and Equipment

1. Hot plate
2. Ice cubes
3. Büchner funnel in No.7 one-hole rubber stopper
4. 500-mL filter flask
5. Filter paper, 7 cm diameter
6. pHydrion paper
7. Boiling chips
8. 95% ethanol
9. Saturated sodium chloride solution
10. 25% NaOH
11. Vegetable oil
12. 5% $FeCl_3$
13. 5% $CaCl_2$
14. Mineral oil
15. 5% $MgCl_2$

NAME ____________ SECTION ______ DATE ______

PARTNER ____________ GRADE ______

Experiment 12

PRE-LAB QUESTIONS

1. If you react vegetable oil with KOH what chemical species will you find at the end of the reaction?

2. Consult Table 10.2 in your textbook. If corn oil is used for making soap what is the chemical formula of the most abundant soap you formed?

3. Indicate which end of the soap molecule is nonpolar and which end is polar.

4. Why is the use of a soap in hard water impractical? Explain with a chemical equation.

NAME ______________________ SECTION __________ DATE __________

PARTNER ______________________ GRADE __________

Experiment 12

REPORT SHEET

Preparation

Appearance of your soap ______________________

Properties

Emulsifying Properties
Which mixture, oil-water or oil-water-soap, forms a more stable emulsion? ______________________

Hard Water Reaction

No. 1 + $CaCl_2$ ______________________

No. 2 + $MgCl_2$ ______________________

No. 3 + $FeCl_3$ ______________________

No. 4 + tap water ______________________

Alkalinity

pH of your soap solution (no. 5) ______________________

Is your soap solution good enough for washing hands?________

POST-LAB QUESTIONS

1. In soap making, first you dissolved vegetable oil in ethanol. What happened to the ethanol during the reaction?

2. What is the major difference between soaps made from lard (animal fat) and from vegetable oil?

3. Explain how soap acts as a cleansing agent?

EXPERIMENT 13

Preparation of an hand cream

Background

Hand creams are formulated to carry out a variety of cosmetic functions. Among these are : softening the skin and preventing dryness; elimination of natural waste products (oils) by emulsification; cooling the skin by radiation, thus helping to maintain body temperature. In addition, hand creams must have certain ingredients that help spreadibility and provide body to the hand cream. In many cases added fragrance improves the odor, and in some special cases medications combat assorted ills.

The basic hand cream formulations all contain water to provide moisture and lanolin to help its absorption by the skin. The latter is a yellowish wax. Chemically wax is made of esters of long chain fatty acids and long chain alcohols. Lanolin is usually obtained from sheep wool; it has the ability to absorb 25-30% of its own weight of water and to form a fine emulsion. Mineral oil, which consists of high molecular weight hydrocarbons, provide spreadibility. In order to allow nonpolar substances, such as lanolin and mineral oil, to be uniformly dispersed in a polar medium, water, one needs strong emulsifying agents. An emulsifying agent must have nonpolar, hydrophobic portions to interact with the oil and also polar, hydrophilic portions to interact with water. A mixture of stearic acid and triethanolamine, through acid-base reaction, yields the salt that has the requirements to act as an emulsifying agent.

Beside these five basic ingredients some hand creams contain also alcohols such as propylene glycol (1,2-propanediol) and esters such as methyl stearate to provide the desired texture of the hand cream.

In this experiment you will prepare four hand creams using combination of ingredients as shown in Table 13.1.

Objectives

1. To learn the method of preparing a hand cream
2. To appraise the function of the ingredients in the hand cream.

Procedure

Preparation of the Hand Creams

For each sample in Table 13.1 assemble the ingredients in two beakers. Beaker 1 contains the polar ingredients and Beaker 2 the nonpolar contents.

Table 13.1

Ingredients	Sample 1	Sample 2	Sample 3	Sample 4	
Water	25 mL	25 mL	25 mL	25 mL	
Triethanolamine	1 mL	1 mL	1 mL	-	Beaker 1
Propylene glycol	0.5 mL	0.5 mL	-	0.5 mL	
Stearic acid	5 g	5 g	5 g	5 g	
Methyl stearate	0.5 g	0.5 g	-	0.5 g	
Lanolin	4 g	4 g	4 g	4 g	Beaker 2
Mineral oil	5 mL	-	5 mL	5 mL	

1. To prepare sample 1 put the nonpolar ingredients in a 50-mL beaker (beaker 2) and heat it in a water bath. The water bath can be a 400-mL beaker half filled with tap water and heated with a Bunsen burner (Fig. 13.1). Carefully hold the beaker with crucible tongues in the boiling water until all ingredients melt.

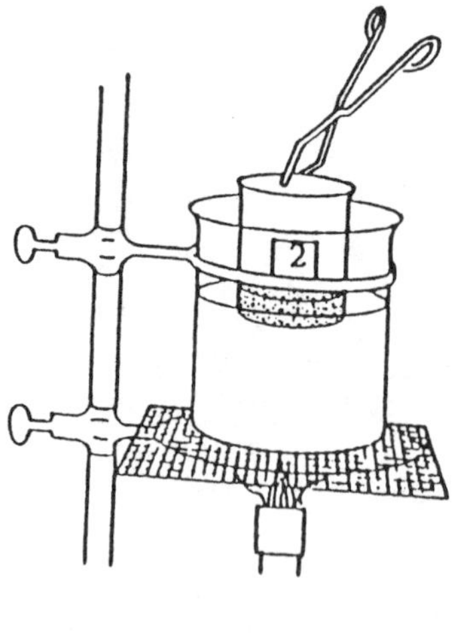

Figure 13.1 Heating ingredients.

2. In the same water bath heat the 100-mL beaker (beaker 1) containing the polar ingredients for about 5 min. Remove the beaker and set it on the bench top.

3. Into the 100-mL beaker containing polar ingredients pour slowly the contents of the 50-mL beaker that holds the molten nonpolar ingredients (Fig. 13.2). Stir the mixture for 5 minutes until you have a smooth uniform paste.

4. Repeat the same procedure in preparing the other three samples.

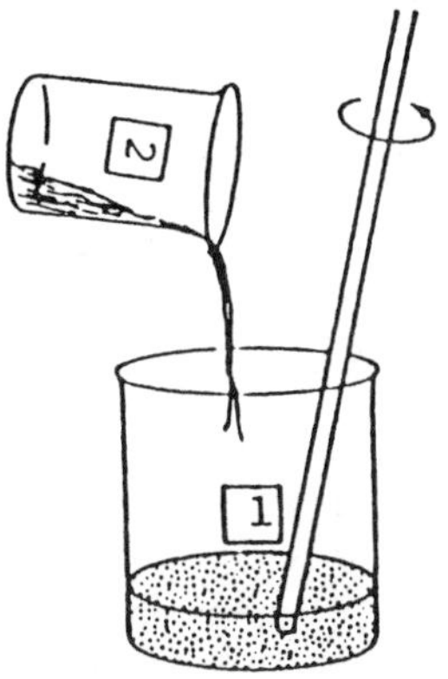

Figure 13.2 Mixing hand cream ingredients.

Characterization of the Hand Cream Preparations

1. Test the pH of the hand creams prepared using a wide range pH paper.

2. Rubbing a small amount of the hand cream between your fingers test for smoothness and homogeneity. Also note the appearance.

Chemicals and Equipment

1. Bunsen burner
2. Lanolin
3. Stearic acid
4. Methyl stearate
5. Mineral oil
6. Triethanolamine
7. Propylene glycol

NAME SECTION DATE

PARTNER GRADE

Experiment 13

PRE-LAB QUESTIONS

1. What is the function of lanolin in the hand cream?

2. The emulsifying agent was prepared from stearic acid and triethanolamine. Give the name of this salt. Write its formula.

3. What functional groups of the emulsifying agent provide the hydrophilic character?

NAME SECTION DATE

PARTNER GRADE

Experiment 13

REPORT SHEET

Characterization of the Hand Cream Samples

Properties	Sample 1	Sample 2	Sample 3	Sample 4
pH				
Smoothness				
Homogeneity				
Appearance				

POST-LAB QUESTIONS

1. In comparing the properties of the hand creams you produced, ascertain what is the function of each of the missing ingredients in the hand cream:

   (a) Mineral oil

   (b) Triethenolamine

   (c) Methyl stearate and propylene glycol

2. A hand cream appears smooth and uniform after you prepared it, but in a week of storage most of the water settles on the bottom and most of the oil separates on the top. What do you think may have gone wrong with the hand cream preparation?

3. Could you prepare a hand cream without water? Would it serve the cosmetic functions listed in the "Background" section?

EXPERIMENT 14

Isolation of lipids from egg yolk

Background

Lipids are a group of chemicals that are characterized by their insolubility in water. The lipids are divided into three classes: simple lipids (fats, oils, and waxes), complex lipids (phospholipids, glycolipids, and sphingolipids), and steroids. These compounds can be separated from each other on the basis of their solubility in different organic solvents. (The fourth class, prostaglandins and leukotrienes, does not fit into this separation scheme).

Egg yolk has a rich supply of lipids. The two most prominent lipids in the egg yolk are cholesterol, which is a steroid, and phosphatidyl choline (or lecithin by its common name), which is a phospholipid.

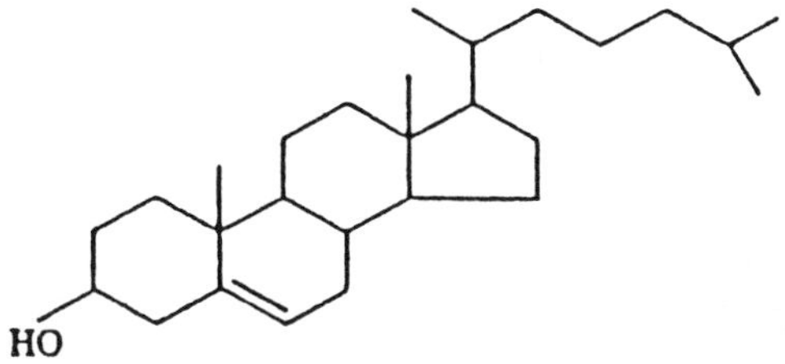

Cholesterol

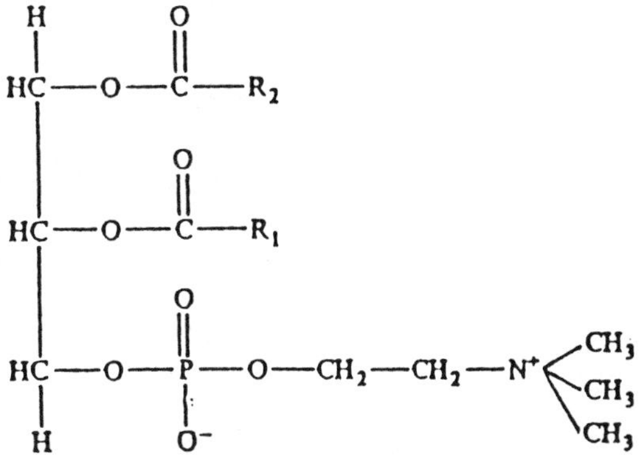

Lecithin

In this experiment, we isolate these two compounds. The egg yolk also contains smaller amounts of neutral fat, glycolipids, and sphingolipids. These may appear as contaminants in your preparations. We do not aim to isolate these minor constituents.

The basis for the isolation of cholesterol and lecithin is their solubility in acetone and ethyl ether. Cholesterol is soluble in acetone, lecithin is not. First, we extract the cholesterol with acetone. Next, we extract the lecithin from the residue of the egg yolk by using ethyl ether as a solvent. Other nutrients (carbohydrates and proteins) are left behind in the residue.

The extracts are not pure compounds. They contain minor lipid constituents. We purify the cholesterol by crystallization from its solution. The contaminant lipid components will stay in solution. We purify the lecithin by precipitating it from the ethyl ether solution. This is accomplished by adding acetone to the solution. The contaminating minor lipid constituents will stay in the ethyl ether-acetone solution. Lecithin and cholesterol are both important constituents of the membranes of cells. They are found in high concentration in the brain and nerves. Cholesterol is also found in the blood. It serves as raw material for the synthesis of many steroids. When the cholesterol concentration increases and the amount of the lipoprotein, the carrier of cholesterol in the blood, is insufficient, cholesterol may be deposited in the blood vessels in the form of plaques (artherosclerosis). Excess cholesterol also forms gallstones.

Lecithin, having a strong polar end and long nonpolar tails of fatty acids, is a good emulsifying agent, dispersing fat in water. The food industry uses lecithin as an emulsifying agent in many products.

Objectives

1. To illustrate the differential solubility of cholesterol and lecithin.
2. To isolate and purify these compounds on the basis of this property alone.

Procedure

CAUTION! Be absolutely certain that no flames (Bunsen burner, matches) are used during this lab period. The solvents employed are flammable and the vapors of the solvents can be ignited by an open flame. Only hot plates are allowed to heat steam baths or, if live steam is available, use that directly for the steam bath. Make certain that when you disconnect the hot plate, you pull the plug and not the cord from the outlet. Pulling the cord occasionally may create sparks, which may ignite the vapors. Keeping your hot plate on your desk bring the water in your steam bath to a boil. Now place your hot steam bath (removed from the hot plate) in the hood and carry out the evaporation of the organic solvents on the steam bath in the hood.

Extraction

1. Take one half the yolk of a hard boiled egg. Record its weight on the Report Sheet (1),(2) and (3). Mash it to a smooth consistency with a spatula in a 250-mL beaker. Add 25 mL acetone. Continue to stir the mixture for 5 min. Allow the solid to settle and decant the acetone extract containing the crude cholesterol. Pour the extract into a labeled 100-mL Erlenmeyer flask and stopper it with a cork. Repeat the extraction with another portion of 25 mL of acetone. Combine the two acetone extracts.

2. Do the next operation under the hood. Evaporate the remaining acetone from the egg yolk residue by putting the beaker on the steam bath. If the evaporation proceeds too fast some of the egg yolk residue may splatter. You can avoid this by removing your beaker occasionaly from the steam bath.The whole evaporation will take a few minutes. Remove the beaker and let it cool. Add 15 mL of ethyl ether (under the hood) to the egg yolk and extract the lecithin by stirring vigorously for 2 min. Decant the ether extract and pour it into a labeled 100-mL Erlenmeyer flask and cork it. Repeat the extraction with another portion of 15-mL ethyl ether. Decant and combine the two extracts. Place the residual egg yolk in a wide mouth jar labeled "Waste."

Isolation and Purification

1. Filter the acetone extract into a clean 100-mL beaker. Evaporate the acetone by placing the beaker on the steam bath in the hood until about 10 mL of extract is left. Transfer this warm acetone extract into a clean and dry 50 mL Erlenmeyer flask. Close the top with a cork. Place the Erlenmeyer flask into an ice bath and wait for 20 min. A white precipitate will form. This is your crude cholesterol preparation.

2. While waiting for the cholesterol to precipitate, you can isolate the lecithin from the ether solution. Weigh a 100-mL beaker. Record the weight on your Report Sheet (4). Filter the combined ether extracts into a preweighed 100-mL beaker. Place the beaker on the steam bath **under the hood** and evaporate most of the ethyl ether until you are left with about 10 mL of ether extract. Cool the beaker to room temperature. Add 30 mL of acetone to the beaker with continuous stirring. A waxy precipitate will form. Decant the solvent and discard it into the jar labeled "Waste." Evaporate the residual solvent by

placing the beaker on the steam bath until the precipitate appears to be dry. As before, during the evaporation,splattering may occur which you can avoid by removing the beaker from the steam bath occasionally. Cool the beaker to room temperature and weigh it. Record it on your Report Sheet (5). Calculate the yield of lecithin and record it on your Report Sheet (6) and (7).

3. Take out the Erlenmeyer flask from the ice bath. Decant most of the cold solvent. Dissolve the cholesterol precipitate in 15 mL of acetone at room temperature. Stir it. Not all the precipitate will dissolve. The contaminating phospholipids will not dissolve. Weigh and record the weight of a watch glass on your Report Sheet (9). Filter the dissolved cholesterol solution into a preweighed watch glass. Allow the acetone to evaporate. The white crystals that formed are your cholesterol preparation. Weigh the watch glass and the cholesterol and record it on your Report Sheet (10). Calculate the yield of the isolated cholesterol and record it on your Report Sheet (11) and (12).

   Preserve both the cholesterol and the lecithin preparations for your next experiment (Exp. 15).

Chemicals and Equipment

1. Hard boiled egg
2. Acetone
3. Ethyl ether
4. Hot plate
5. Steam bath
6. Waste jar

NAME ______________________ SECTION __________ DATE __________

PARTNER ______________________ GRADE ______________________

Experiment 14

PRE-LAB QUESTIONS

1. Lecithin is the name of a group of complex lipids. Write the structure of two different lecithin molecules

2. Acetone can be mixed with water in any proportion. Ethyl ether is insoluble in water. Which one is a more polar solvent, acetone or ethyl ether? Explain.

3. Cholesterol can form well defined crystals in gallstones. What kind of organization and intermolecular interactions would provide the most stable structure of cholesterol in the crystals.

4. Why are you not allowed to use an open flame (Bunsen burner) during this experiment?

NAME ______________________ SECTION __________ DATE __________

PARTNER ______________________ GRADE __________

Experiment 14

REPORT SHEET

1. Weight of the paper (tare) ______________g
2. Weight of the egg yolk + paper ______________g
3. Weight of the egg yolk ______________g
4. Weight of beaker ______________g
5. Weight of lecithin + beaker ______________g
6. Weight of lecithin ______________g
7. % yield of lecithin = (weight of lecithin/ weight of yolk) x 100 = ______________%
8. Appearance of lecithin: ______________
9. Weight of watch glass ______________g
10. Weight of watch glass + cholesterol ______________g
11. Weight of cholesterol ______________g
12. % yield of cholesterol ______________%
13. Appearance of cholesterol: ______________

POST-LAB QUESTIONS

1. Egg yolk is used as an emulsifying agent in many foods (mayonnaise). Which of the isolated components of the egg yolk can perform such a function best? Why?

2. Consumer charts indicate that one egg yolk contains 250 mg of cholesterol. Compare this to your yield. Assuming that your preparation is pure cholesterol, what percent of the available cholesterol did you isolate?

3. The normal cholesterol level in the blood is 1.0 mg/ mL. Your body contains 5.5 L of blood. You just ate two hard boiled eggs. Assuming that all the cholesterol from the yolk is absorbed into your bloodstream, calculate what will be the cholesterol concentration in your blood after the meal.

4. Why does cholesterol, in the absence of protective lipoproteins, form plaques in the blood vessels?

EXPERIMENT 15

Analysis of lipids

Background

Lipids are chemically heterogeneous mixtures. The only common property they have is their insolubility in water. We can test for the presence of various lipids by analyzing their chemical constituents. In Experiment 14 you isolated the lipids cholesterol and lecithin from egg yolk. These may be pure compounds or they still may contain other lipids as contaminants.

Structurally, cholesterol is completely different from lecithin. It contains a five-membered fused ring that is the common core of all steroids.

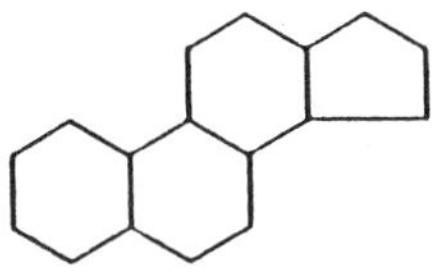

steroid nucleus

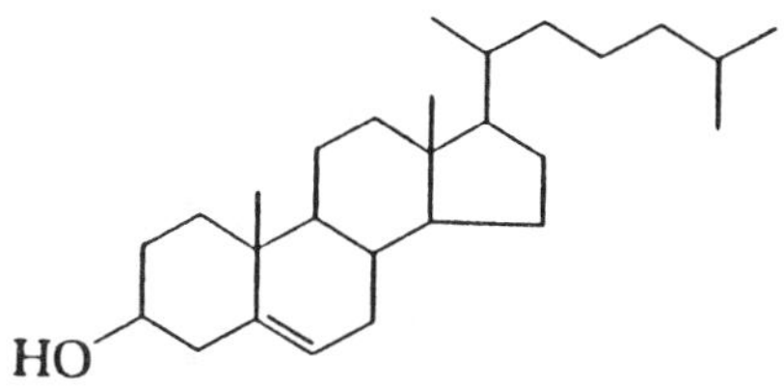

cholesterol

There is a special colorimetric test, the Lieberman-Burchard reaction, which uses acetic anhydride and sulfuric acid as reagents, that gives a characteristic green color in the presence of cholesterol. This color is due to the -OH group of cholesterol and the unsaturation found in the adjacent fused ring. The color change is gradual: first it appears as a pink coloration, changing later to lilac, and finally to deep green.

Lecithin is a complex lipid; it contains a number of components that can be detected and thereby shows the presence of this phospholipid. When lecithin is hydrolyzed in acidic medium, both the fatty-acid ester bonds and the phosphate ester bonds are broken and free fatty acids and inorganic phosphate are released.

$$\text{H}_2\text{C}-\text{O}-\overset{\overset{\text{O}}{\|}}{\text{C}}-(\text{CH}_2)_n\text{CH}_3$$

$$\text{HC}-\text{O}-\overset{\overset{\text{O}}{\|}}{\text{C}}-(\text{CH}_2)_n\text{CH}_3$$

$$\text{H}_2\text{C}-\text{O}-\overset{\overset{\text{O}}{\|}}{\underset{\underset{\text{O}^-}{|}}{\text{P}}}-\text{O}-\text{CH}_2\text{CH}_2\text{N}^+\overset{\text{CH}_3}{\underset{\text{CH}_3}{\diagdown}}\text{CH}_3$$

$$+ 4\text{H}_2\text{O} \overset{\text{H}^+}{\rightleftarrows} \overset{\text{CH}_2\text{OH}}{\underset{\text{CH}_2\text{OH}}{\overset{|}{\text{CHOH}}}} + 2\text{CH}_3(\text{CH}_2)_n\overset{\overset{\text{O}}{\|}}{\text{C}}-\text{OH} + \text{HO}-\overset{\overset{\text{O}}{\|}}{\underset{\underset{\text{O}^-}{|}}{\text{P}}}-\text{OH} + \overset{\text{CH}_2\text{OH}}{\underset{\underset{\text{CH}_3}{\overset{\text{CH}_3}{\diagup}\text{N}^+\overset{\diagdown}{}\text{CH}_3}}{\overset{|}{\text{CH}_2}}}$$

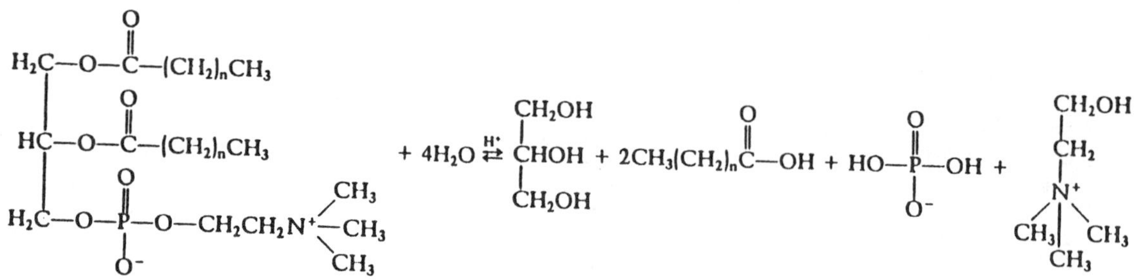

Using a molybdate test, we can detect the presence of phosphate in the hydrolysate by the appearance of a purple color. Although this test is not specific for lecithin (other phosphate containing lipids will give a positive molybdate test), it differentiates clearly between cholesterol (negative test) and phospholipids (positive test).

A second test that differentiates between cholesterol and lecithin is the acrolein reaction. When lipids containing glycerol are heated in the presence of potassium hydrogen sulfate the glycerol is dehydrated, forming acrolein, which has an unpleasant odor. Further heating results in polymerization of acrolein, which is indicated by the slight blackening of the reaction mixture. Both the pungent smell and the black color indicate the presence of glycerol and thereby lecithin. Cholesterol gives a negative acrolein test.

$$\overset{\text{CH}_2\text{OH}}{\underset{\text{CH}_2\text{OH}}{\overset{|}{\text{CHOH}}}} \overset{\Delta}{\rightarrow} \overset{\text{CH}_2}{\underset{\underset{\text{H}}{\overset{|}{\text{C}=\text{O}}}}{\overset{\|}{\text{CH}}}} + 2\text{H}_2\text{O}$$

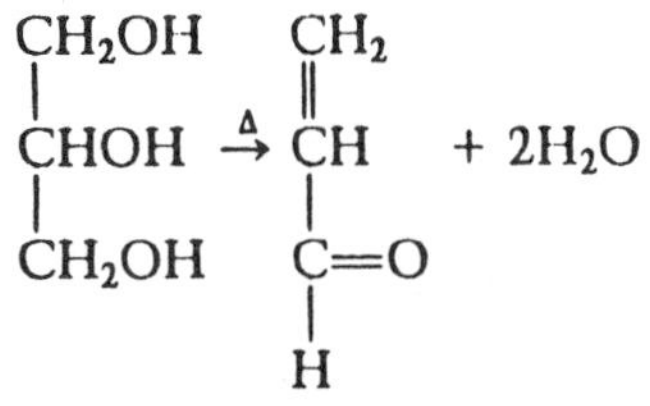

The purity of both the lecithin and cholesterol preparations can be tested by melting point measurements. A pure compound gives a sharp melting point within ±1°C range. Compounds that are contaminated by the presence of other constituents give a relatively broad melting point range of ±5°C. Impurities usually lower the melting point of a compound. Impure lecithin may even decompose before reaching the melting point. By taking the melting points of your cholesterol and lecithin preparations and comparing them with the literature values of the pure compounds, you can get a good idea regarding the purity of your compounds.

Objectives

To investigate the purity of cholesterol and lecithin preparations by using colorimetric tests and melting point measurements.

Procedure

In Experiment 14 you isolated two compounds from egg yolk: cholesterol and lecithin. We will test these compounds for their purity.

Phosphate test

1. Add about 0.2 g of cholesterol to one clean and dry test tube and the same amount of lecithin to another. Hydrolyze the compounds by adding 3 mL of 6 M HNO_3 to each test tube.

CAUTION! 6 M nitric acid is a strong acid. Handle it with care.

2. Prepare a water bath by boiling about 100 mL of tap water in a 250-mL beaker on a hot plate. Place the test tubes in the boiling water bath for 5 min. Do not inhale the vapors. Cool the test tubes. Neutralize the acid by adding 3 mL of 6 M NaOH. Mix.

3. Transfer 2 mL of each neutralized sample into clean and labeled test tubes. Add 3 mL of a molybdate solution to each test tube and mix the contents. **(Be careful. The molybdate solution contains sulfuric acid.)** Heat the test tubes in a boiling water bath for 5 min. Cool them to room temperature.

4. Add 0.5 mL of an ascorbic acid solution and mix the contents thoroughly. Wait for 20 min for the development of the purple color. While you wait you can perform the rest of the colorimetric tests.

The acrolein test for glycerol

1. Place 1 g $KHSO_4$ in a clean, dry test tube and add a few crystals of your cholesterol preparation.

2. To a second test tube, add also 1 g of $KHS0_4$ and a few grains of your lecithin preparation. Set up your Bunsen burner in the hood. It is important that this test should be performed under the hood because of the pungent odor of the acrolein.

3. Gently heat each test tube, one at a time, over the Bunsen burner's flame, shaking it continuously from side to side. When the mixture melts, it slightly blackens and you will notice the evolution of fumes. Stop the heating. Smell the test tubes by moving them sideways under your nose or waft the vapors. Do not inhale the fumes directly. A pungent odor, resembling burnt hamburgers, is the positive test for glycerol. Do not overheat the test tubes, for the residue will become hard, making it difficult to clean the test tubes.

Lieberman-Burchard test for cholesterol

Place a few crystals of your cholesterol preparation in one labeled test tube. Add about the same amount of lecithin to a second clean and labeled test tube. **(The next step should be done in the hood).** Transfer 3 mL of chloroform and 1 mL of acetic anhydride to each test tube. Finally, add one drop of concentrated sulfuric acid to each mixture. Mix the contents and record the color changes, if any. Wait 5 min. Record again the color of your solutions.

Melting point measurements

1. Place a few grains of cholesterol crystals into a capillary tube. Bring the crystals to the closed bottom end of the capillary by either using a file in a bowing motion across the capillary or dropping the capillary repeatedly through a 30 to 40 cm glass tube.

2. Place the capillary tube in a melting point measuring apparatus. Record the melting point of your cholesterol preparation. The CRC Handbook of Chemistry and Physics lists a value for the melting point of cholesterol as 148°C. Compare your result.

3. Repeat the melting point determination with your lecithin preparation. If your lecithin is waxlike, you may have to disperse your sample to a powdery consistency in a mortar with pestle before you can place it in your capillary. The Handbook gives the melting point of lecithin as 236°C.

Chemicals and Equipment

1. 6 M NaOH
2. 6 M HNO_3
3. Molybdate reagent
4. Ascorbic acid solution
5. Potassium hydrogen sulfate, $KHSO_4$
6. Chloroform
7. Acetic anhydride
8. Sulfuric acid, H_2SO_4
9. Melting point apparatus
10. Capillary tubes
11. Hot plate

NAME SECTION DATE

PARTNER GRADE

Experiment 15

PRE-LAB QUESTIONS

1. What is the difference in intermolecular interaction in lecithins vs. cholesterol that gives such a wide contrast in their melting points?

2. Cholesterol has an alcohol group. One could also dehydrate cholesterol (removing one water molecule by heating). Show what kind of structure you would expect from the dehydration of cholesterol.

3. List all the functional groups in (a) acrolein (b) choline.

4. What techniques can you use to get the cholesterol crystals to the bottom of the capillary tube?

POST-LAB QUESTIONS

1. What is your overall conclusion regarding the purity of your compounds?

2. Assume that your cholesterol was free from all other lipid contaminants. However, it was not completely dry; it still contained some acetone. How would that affect your tests?

   a. Acrolein:

   b. Lieberman-Burchard:

   c. Phosphate:

   d. Melting point:

3. Describe in detail the appearance of your cholesterol before and after you reached the melting point.

4. The thermometer in your melting point apparatus is not properly calibrated. It reads 102°C in boiling water. How does that affect your melting point measurements of cholesterol and lecithin?

NAME SECTION DATE

PARTNER GRADE

Experiment 15

REPORT SHEET

Tests	Cholesterol	Lecithin
1. Phosphate		
a. Color		
b. Conclusions		
2. Acrolein		
a. Odor		
b. Color		
c. Conclusions		
3. Lieberman-Burchard		
a. Initial color		
b. Color after 5 min		
c. Conclusion		
4. Melting point		
Comments		

POST-LAB QUESTIONS

1. What is your overall conclusion regarding the purity of your compounds?

2. Assume that your cholesterol was free from all other lipid contaminants. However, it was not completely dry; it still contained some acetone. How would that affect your tests?

   a. Acrolein:

   b. Lieberman-Burchard:

   c. Phosphate:

   d. Melting point:

3. Assume that the lecithin still contained all its ester linkages. Would you get a positive acrolein test? Explain.

4. A positive acrolein test is indicated by its odor as well as by its color. Which comes first? Explain.

EXPERIMENT 16
TLC separation of amino acids

Background

Amino acids are the building blocks of peptides and proteins. They possess two functional groups: the carboxyl group gives the acidic character, and the amino group provides the basic character. The common structure of all amino acids is

$$R-\underset{\underset{NH_2}{|}}{\overset{\overset{H}{|}}{C}}-COOH$$

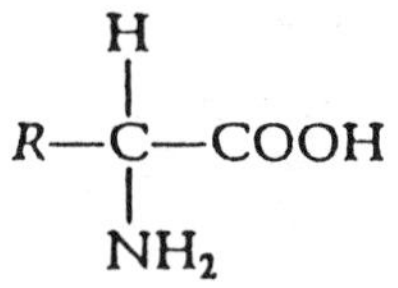

The R represent the side chain that is different for each of the amino acids that are commonly found in proteins. However, all 20 amino acids have a free carboxyl group and a free amino (primary amine) group, except proline which has a cyclic side chain and a secondary amino group.

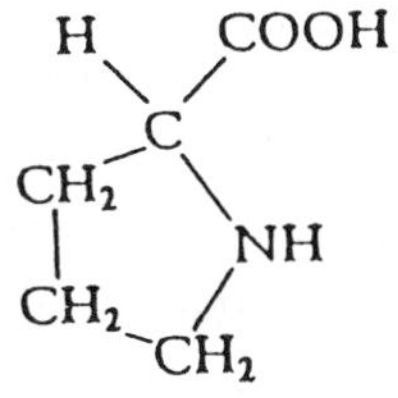

proline

We use the properties provided by these groups to characterize the amino acids. The common carboxyl and amino groups provide the acid-base nature of the amino acids. The different side chains, and the solubilities provided by these side chains can be utilized to identify the different amino acids by their rate of migration in thin layer chromatography.

In this experiment we use thin layer chromatography to identify aspartame, an artificial sweetener, and its hydrolysis products in certain foods.

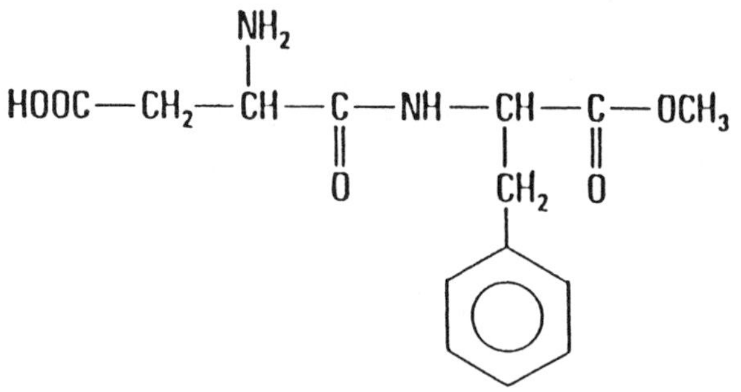

Aspartame is the methyl ester of the dipeptide aspartylphenylalanine. Upon hydrolysis with HCl it yields aspartic acid, phenylalanine and methyl alcohol. When this artificial sweetener was approved by the Food and Drug Administration, opponents of aspartame claimed that it is a health hazard, because in soft drinks upon long storage aspartame would be hydrolyzed and would yield poisonous methyl alcohol. The Food and Drug Administration ruled, however, that aspartame is sufficiently stable and fit for human consumption. Only a warning must be put on the labels containing aspartame. This warning is for patients suffering from phenylketonurea who cannot tolerate phenylalanine.

To run a thin layer chromatography experiment we use silica gel in a thin layer on a plastic or on a glass plate. We apply the sample (aspartame or amino acids) as a spot to a strip of a thin layer plate. The plate is dipped into a mixture of solvents. The solvent moves up the thin gel by capillary action and carries the sample with it. Each amino acid may have a different migration rate depending on the solubility of the side chain in the solvent. Amino acids with similar side chains are expected to move with similar though not identical rates; those that have quite different side chains are expected to migrate with different velocities. Depending on the solvent system used, almost all amino acids and dipeptides can be separated from each other by thin layer chromatography (TLC).

We actually do not measure the rate of migration of an amino acid or a dipeptide, but rather, how far a particular amino acid travels in the thin silica gel layer relative to the migration of the solvent. This ratio is called the R_f value. In order to calculate the R_f values, one must be able to visualize the position of the amino acid or dipeptide. This is done by spraying the thin layer silica gel plate with a ninhydrin solution that reacts with the amino group of the amino acid. A purple color is produced when the gel is heated. (The proline not having a primary amine gives a yellow color with ninhydrin). For example, if the purple spot of an amino acid appears on the TLC 4.5 cm away from the origin and the solvent front migrates 9.0 cm (Fig.16.1) the R_f value for the amino acid is

$$R_f = \frac{\text{distance traveled by the amino acid}}{\text{distance traveled by the solvent front}} = \frac{4.5\text{ cm}}{9.0\text{ cm}} = 0.50$$

In the present experiment you will determine the R_f values of three amino acids, phenylalanine, aspartic acid and leucine. You will also measure the R_f value of aspartame.

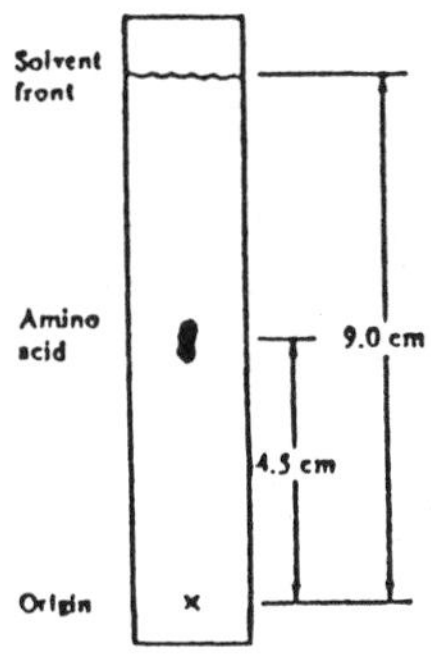

Figure 16.1 TLC chromatogram

The aspartame you will analyze is actually a commercial sweetener, Equal by NutraSweet Co., that contains besides aspartame, silicon dioxide, glucose, cellulose and calcium phosphate. None of these other ingredients of Equal will give a purple or any other colored spot with ninhydrin. Occasionally some sweeteners may contain a small amount of leucine which can be detected by the ninhydrin test. You will also hydrolyze aspartame using HCl as a catalyst to see if the hydrolysis products will prove that the sweetener was truly aspartame. Finally you will analyze some commercial soft drinks, supplied by your instructor.

The analysis of the soft drink can tell you if the aspartame was hydrolyzed at all during the processing and storing of the soft drink.

Objectives

1. To separate amino acids and a dipeptide by TLC.
2. To identify hydrolysis products of aspartame.
3. To analyze the state of aspartame in soft drinks.

Procedure

1. Dissolve about 10 mg of the sweetener Equal in 1 mL of 3 M HCl in a test tube. Heat it with a Bunsen burner to boil for 30 seconds, but make sure the liquid should not completely evaporate. Cool the test tube and label it as "Hydrolyzed Aspartame".

2. Label 5 additional small test tubes, respectively, for aspartic acid, phenylalanine, leucine, aspartame and Diet Coca-Cola.
   Place about 0.5 mL samples in each test tube.

3. Take two 15 x 6.5 cm TLC plates. With a pencil lightly draw a line parallel to the 6.5 cm edge and about 1 cm from the edge. Mark the positions of 5 spots on each plate, paced equally, where you will spot your samples (Fig. 16.2). You must make sure that you don't touch the plates with your fingers.

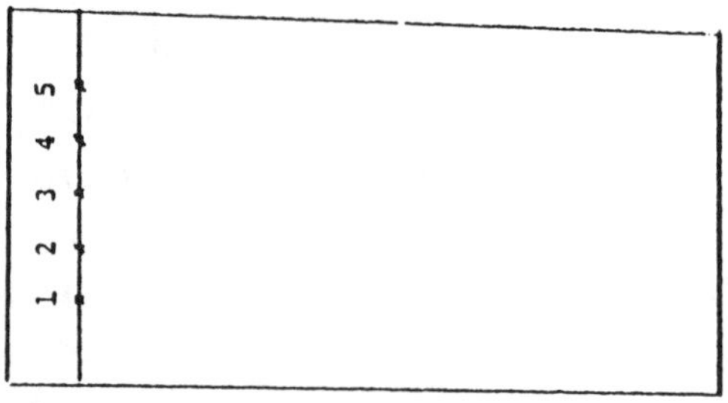

Figure 16.2 Spotting

Either use plastic gloves or handle the plates by holding them only at their edges. This precaution must be observed throughout the whole operation, because amino acids from your fingers will contaminate the plate.
On plate A you will spot samples of (1) phenylalanine,(2) aspartic acid, (3) leucine, (4) aspartame in Equal and (5) hydrolized aspartame you prepared in step 1. On plate B you will spot samples of Diet Coca-Cola on lanes (1) and (4), aspartic acid on lane (2), aspartame in Equal on lane (3), and the hydrolyzed aspartame you prepared previously on lane (5).

4. First spot plate A. For each sample use a separate capillary tube. Apply the sample to the plate until it spreads to a spot of 1 mm diameter. Dry the spots. (If a heat lamp is available, use it for drying). Pour about 15 mL of solvent mixture (butanol:acetic acid:water) into a large (500 mL or 1 L) beaker and place your spotted plate diagonally for ascending chromatography. Make certain that the spots applied to the plate are above the surface of the eluting solvent. Cover the beaker with aluminum foil to avoid the evaporation of the solvent mixture.

5. Spot plate B. For aspartic acid, lane (2), and for the hydrolyzed and non-hydrolyzed aspartame, lanes (3) and (5), use one spot as before. For Coca-Cola (lanes (1) and (4)) multiple spotting is needed. Apply the capillary tube 12-15 times to the same spot, making certain that between each application the previous sample has been dried. Also try to control the size of the spots that they should not spread too much, not more than 2 mm in diameter. Dry the spots as before. Place the plate in a large beaker containing the eluting solvent as before. Cover the beaker with aluminum foil. Allow about 50-60 minutes for the solvent front to advance.

6. When the solvent front nears the edge of the plate, about 1-2 cm from the edge, remove the plate from the beaker. You must not allow the solvent front to advance up to or beyond the edge of the plate. Mark immediatly with a pencil the position of the solvent front. Under a hood dry the plates with the aid of a heat lamp or hair dryer. Using polyethylene gloves spray the dry plates with ninhydrin solution. Be careful not to spray ninhydrin on your hand and not to touch the sprayed areas with bare hands. If the ninhydrin spray touches your skin which contains amino acids, your fingers will be discolored for a few days. Place the sprayed plates into a drying oven at 105-110°C for 2-3 minutes.

7. Remove the plates from the oven. Mark the center of the spots and calculate the R_f values of each spot.

Chemicals and Equipment

1. 0.1% solutions of aspartic acid, phenylalanine and leucine
2. 0.5% solution of aspartame (Equal)
3. Diet Coca-Cola
4. 3 M HCl
5. 0.2% ninhydrine spray
6. Butanol: acetic acid: water solvent mixture
7. Equal sweetener
8. Aluminum foil
9. 15 x 6.5 cm silica gel TLC plates
10. Ruler
11. Polyethylene gloves
12. Capillary tubes open on both ends
13. Heat lamp or hair dryer
14. Drying oven ,110°C

NAME SECTION DATE

PARTNER GRADE

Experiment 16

PRE-LAB QUESTIONS

1. If an amino acid moved 2.1 cm on a TLC plate and the solvent moved 7.0 cm, what is the R_f value of the amino acid?

2. Why must you use a pencil and not ink to mark the origin of your spot on the TLC plate?

3. What would happen if you didn't use gloves and your finger comes in contact with the ninhydrin spray?

4. How would you calculate the R_f values of your samples if you would allow the solvent front to run over the edges of your TLC plates?

NAME SECTION DATE

PARTNER GRADE

Experiment 16

REPORT SHEET

1. Sample	Distance traveled (mm)	Solvent front (mm)	R_f
Aspartic acid			
Phenylalanine			
Leucine			
Aspartame			
Hydrolyzed aspartame			
Diet Coca-Cola			

2. Identification
   (a) Name the amino acids you found in the hydrolysate of the sweetener, Equal.

   (b) How many spots were stained with ninhydrin (1) in Equal and (2) in Coca-Cola samples?

POST-LAB QUESTIONS

1. Can you separate aspartame, a dipeptide, from its constituent amino acids by TLC technique? Which sample migrated the slowest?

2. In testing the hydrolysate of aspartame, you forgot to mark the position of the solvent front on your TLC plate. Could you
(a) determine how many amino acids were in the aspartame;
(b) identify those amino acids?

3. Do you have any evidence that the aspartame was hydrolyzed during the processing and storage of the Diet Coca-Cola sample? Explain.

4. How could you have achieved a better separation on your TLC plate between aspartic acid and phenylalanine?

5. If the R_f values of two amino acids are 0.3 and 0.35, respectively, what length should the TLC plate be in order that the two spots should be separated by 1 cm?

EXPERIMENT 17
Acid-base properties of amino acids

Background

In the body, amino acids exist as zwitterions

$$R-\overset{\displaystyle H}{\underset{\displaystyle NH_3^+}{C}}-COO^-$$

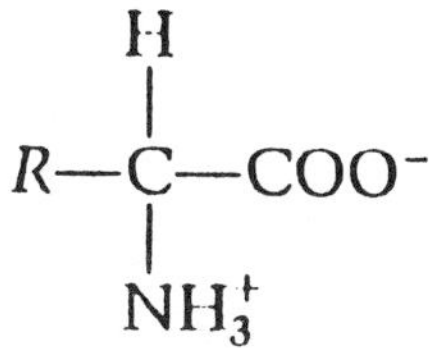

This is an amphoteric compound because it behaves as both an acid and a base in the Bronsted definition. As an acid it can donate a H^+ and becomes the conjugate base:

$$R-\overset{\displaystyle H}{\underset{\displaystyle NH_3^+}{C}}-COO^- + OH^- \rightleftarrows R-\overset{\displaystyle H}{\underset{\displaystyle NH_2}{C}}-COO^- + H_2O$$

$$\text{acid} \qquad \text{base} \qquad \text{conj. base} \qquad \text{conj. acid}$$

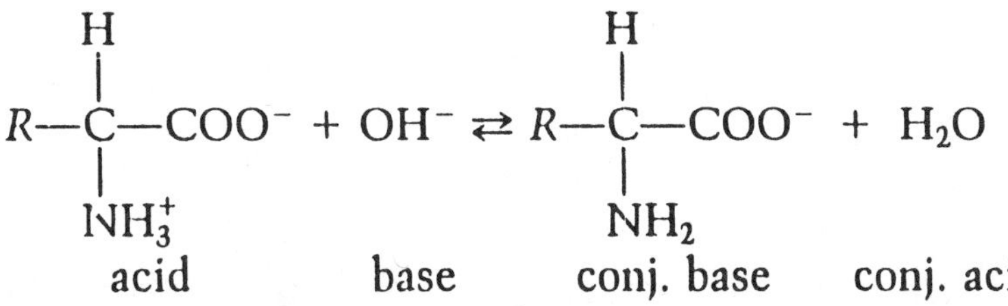

As a base it can accept a H^+ ion and will become a conjugate acid:

$$R-\overset{\displaystyle H}{\underset{\displaystyle NH_3^+}{C}}-COO^- + H_3O^+ \rightleftarrows R-\overset{\displaystyle H}{\underset{\displaystyle NH_3^+}{C}}-COOH + H_2O$$

$$\text{base} \qquad \text{acid} \qquad \text{conj. acid} \qquad \text{conj. base}$$

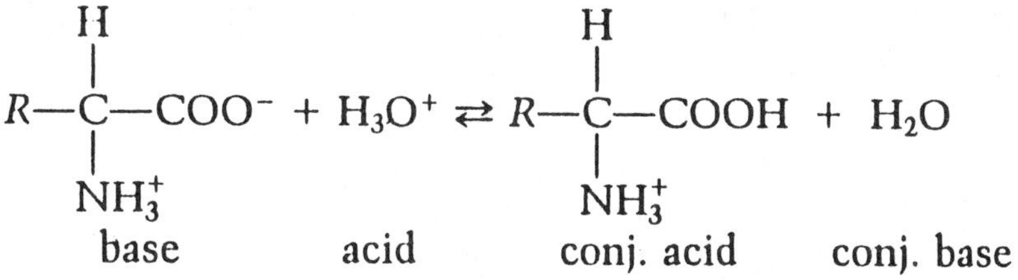

To study the acid-base properties one can perform a simple titration. We start our titration with the amino acid being in its acidic form at a low pH:

$$R-\overset{\displaystyle H}{\underset{\displaystyle NH_3^+}{C}}-COOH \quad (I)$$

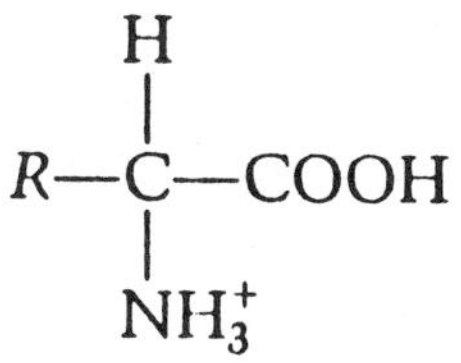

As we add a base, OH⁻, to the solution, the pH will rise. We record the pH of the solution by a pH meter after each addition of the base. To obtain the titration curve, we plot the milliliters of NaOH added against the pH of the solution (Fig 17.1). Note that there are two flat portions (called legs) on the titration curve where the pH does not increase appreciably

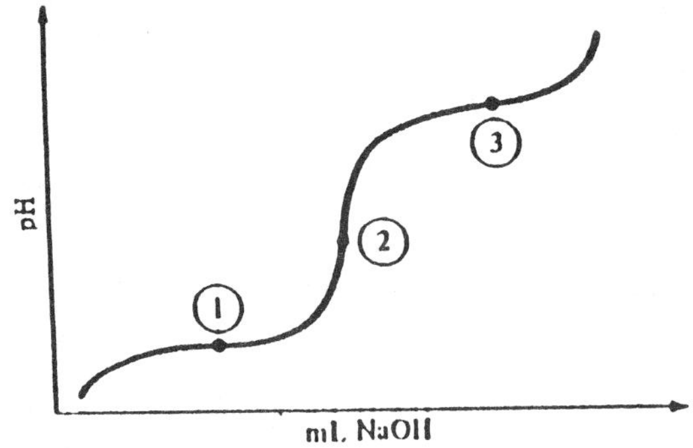

Figure 17.1 The titration curve of an amino acid.

with the addition of NaOH. The midpoint of the first leg, 1, is when half of the original acidic amino acid (I) has been titrated and it became a zwitterion.

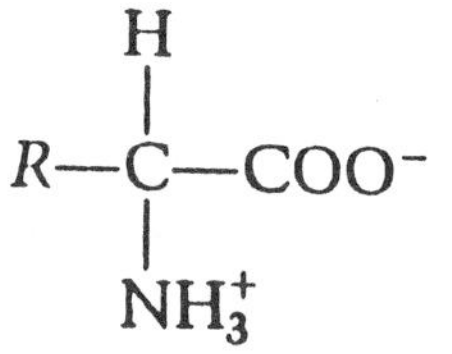

$$\begin{array}{c} H \\ | \\ R-C-COO^- \quad (II) \\ | \\ NH_3^+ \end{array}$$

The point of inflection, 2, occurs when the amino acid is entirely in the zwitterion form (II). At the midpoint of the second leg, 3, half of the amino acid is in the zwitterion form and half is in the basic form (III).

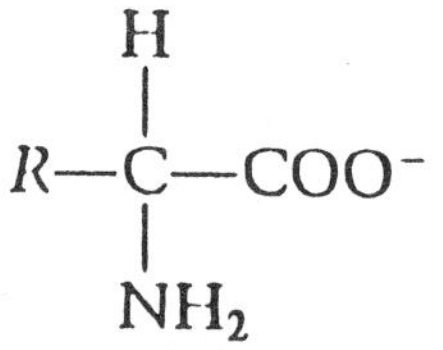

$$\begin{array}{c} H \\ | \\ R-C-COO^- \quad (III) \\ | \\ NH_2 \end{array}$$

From the pH at the midpoint of the first leg we obtain the pK value of the carboxyl group, since this is the group that is titrated with NaOH at this stage (the structure going from I to II). The pH of the midpoint of the second leg, 3, is equal to the pK of the -NH₃⁺, since this is the functional group that donates its H⁺ at this stage of the titration. The pH at the inflection point, 2, is equal to the isoelectric point. At the isoelectric point of a compound the positive and negative charges balance each other. This occurs at the inflection point when all the amino acids are in the zwitterion form.

You will obtain a titration curve of an amino acid with a neutral side chain such as glycine, alanine, phenylalanine, leucine or valine. If pH meters are available you read the pH directly from the instrument after each addition of the base. If a pH meter is not available, you can obtain the pH with the aid of indicator papers. From the titration curve obtained you can read the pK values and the isoelectric point.

Objectives

1. To study acid base properties by titration.
2. To calculate pK values for the titratable groups.

Procedure

1. Pipet 20 mL of 0.1 M an amino acid solution (glycine, alanine, phenylalanine, leucine or valine) that has been acidified with HCl to a pH of 1.5 into a 100 mL beaker.
2. If a pH meter is available, insert the clean and dry electrode of the pH meter into a standard buffer solution with known pH. Turn the knob of the meter to the pH mark and adjust it to read the pH of the buffer. Turn the knob of the pH meter to "Standby" position. Remove the electrode from the buffer, wash it with distilled water and dry it. Insert the dry electrode into the amino acid solution. Turn the knob of the meter to "pH" position and record the pH of the solution.

   Fill a buret with 0.25 M NaOH solution. Add the NaOH solution from the buret in 1.0 mL increments to the beaker. After each increment, stir the contents with a glass rod and then read the pH of the solution. Continue the titration as described until you reach pH 12.

   Turn off your pH meter and wash the electrode with distilled water, wipe it dry and store it in its original buffer.
3. If a pH meter is not available, perform the titration as above but use pH indicator papers. After the addition of each increment and stirring, withdraw a drop of the solution with a Pasteur pipet. Touch the end of the pipet to a dry piece of the pH indicator paper. Compare the color of the indicator paper with the color on the charts supplied. Read the corresponding pH from the chart and record it.
4. Draw your titration curve. From the graph, read your pK values and the isoelectric point of your amino acid.

Chemicals and Equipment

1. 0.1 M amino acid solution (glycine, alanine, leucine, phenylalanine or valine)
2. 0.25 M NaOH solution
3. pH meter and standard buffer (or pH indicator paper and Pasteur pipet)
4. 50 mL buret
5. 20 mL pipet
6. Spectroline pipet filler

NAME ______ SECTION ______ DATE ______

PARTNER ______ GRADE ______

Experiment 17

PRE-LAB QUESTIONS

1. Write the acidic, amphoteric (zwitterion) and the basic form of leucine.

2. If the equilibrium constant, K_a, for the ionization of the carboxyl group is 1×10^{-4}, what is the pK_a?

3. If in a solution of alanine, the number of negatively charged carboxyl groups, $-COO^-$, is 1×10^{-19} at the isoelectric point, what is the number of the positively charged amino groups, $-NH_3^+$?

NAME ______ SECTION ______ DATE ______

PARTNER ______ GRADE ______

Experiment 17

REPORT SHEET

1. Amino acid used for titration ______________________

mL of 0.25 M NaOH added	pH
0	
1.0	
2.0	
3.0	
4.0	
5.0	
6.0	
7.0	
8.0	
9.0	
10.0	
11.0	
12.0	

2. Plot your data below to get the titration curve.

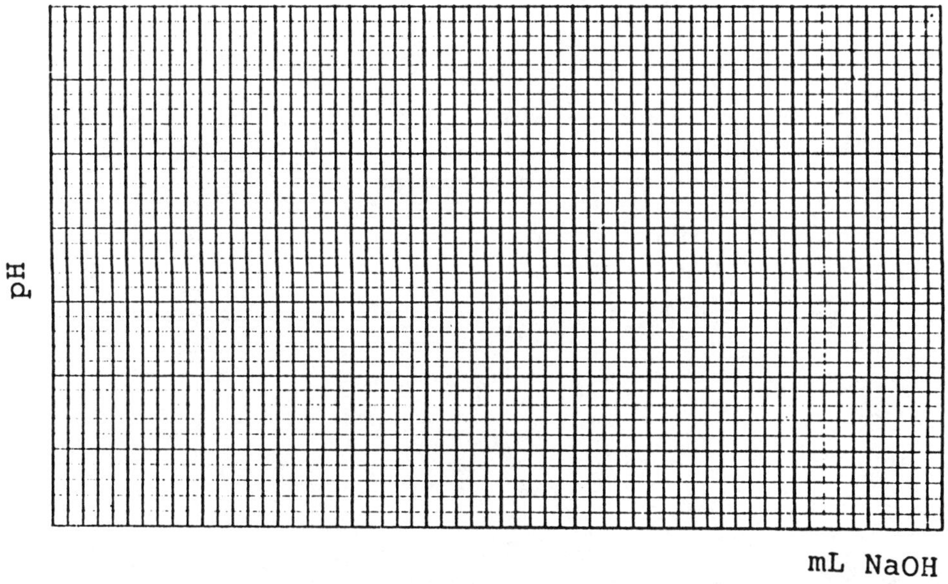

3. a. Indicate the positions of the midpoints of each leg and position of the inflection point your graph.
   b. Record the pK values for the carboxyl group____, for the amino group__________.
   c. Record the pH of the isoelectric point__________.

POST-LAB QUESTIONS

1. The isoelectric point of an amino acid is an intensive property.
   (a) Knowing that, would you expect to find your inflection point at a different pH value, if you had titrated 0.5 M solution of the same amino acid instead of the 0.1 M solution?

   (b) Would your result be different if you had used 50 mL of amino acid solution instead of 20 mL?

2. Write the structures of the amino acids that exist in your solutions at pH 4 and at pH 12.

3. Which data can you obtain with greater accuracy from your graph, the pK values or the isoelectric point? Explain.

EXPERIMENT 18

Isolation and identification of casein.

Background

Casein is the most important protein in milk. It functions as a storage protein fulfilling nutritional requirements. Casein can be isolated from milk by acidification to bring it to its isoelectric point. At the isoelectric point the number of positive charges on a protein equal the number of negative charges. Proteins are least soluble in water at their isoelectric points because they tend to aggregate by electrostatic interaction. The positive end of one protein molecule attracts the negative end of another protein molecule and the aggregates precipitate out of solution.

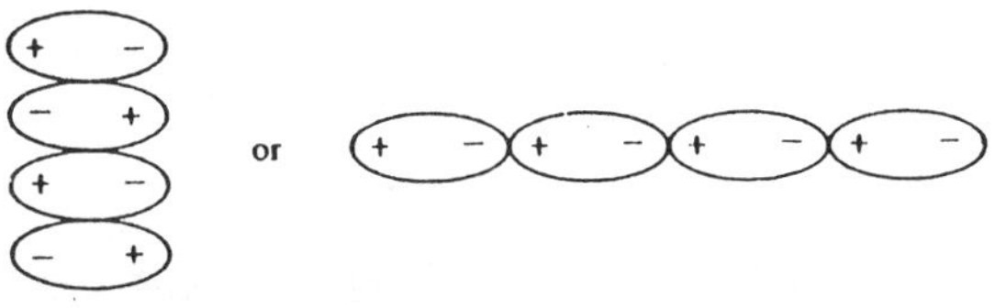

On the other hand, if a protein molecule has a net positive (at low pH or acidic condition) or a net negative charge (at high pH or basic condition), its solubility in water is increased.

$$\overset{+}{N}H_3\sim COOH \xleftarrow[\text{low pH}]{H^+} \overset{+}{N}H_3\sim COO^- \xrightarrow[\text{high pH}]{OH^+} NH_2\sim COO^- + H_2O$$

more soluble — least soluble (at isoelectric pH) — more soluble

In the first part of this experiment you are going to isolate casein from milk which has a pH of about 7. Casein will be separated as an insoluble precipitate by acidification of the milk to its isoelectric point (pH= 4.6). The fat that precipitates along with casein can be removed by dissolving it in alcohol.

In the second part of this experiment you are going to prove that the precipitated milk product is a protein. The identification will be achieved by performing a few important chemical tests.

1. <u>The Biuret Test.</u> This is one of the most general tests for proteins. When a protein reacts with copper(II) sulfate, a positive test is the formation of a copper complex which has a violet color.

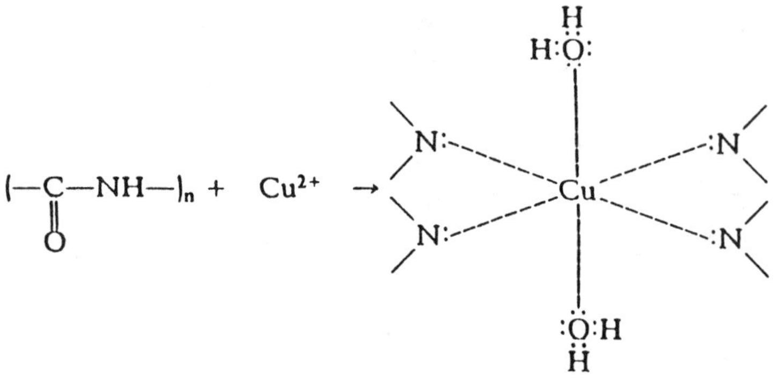

protein	blue color	protein-copper complex
		(violet color)

This test works for any protein or compound that contains two or more of the following groups:

$$-\underset{\underset{O}{\|}}{C}-NH-, \quad -\underset{\underset{O}{\|}}{C}-NH_2, \quad -CH_2-NH_2, \quad -\underset{\underset{NH}{\|}}{C}-NH_2, \quad -\underset{\underset{S}{\|}}{C}-NH_2$$

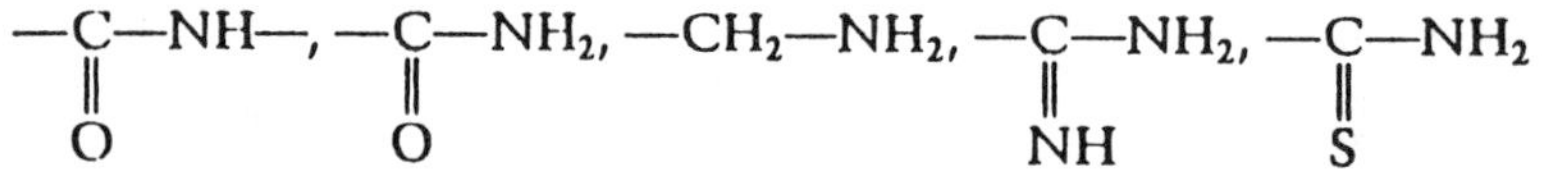

2. <u>The Ninhydrin Test.</u> Amino acids with a free $-NH_2$ group and proteins containing free amino groups react with ninhydrin to give a purple-blue complex.

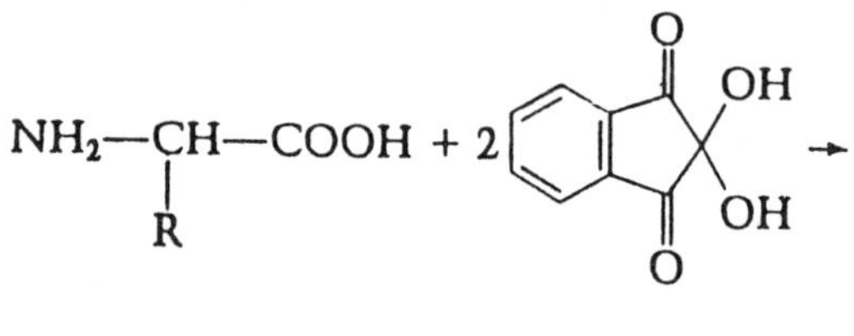

protein	ninhydrin

$+ RCHO + CO_2 + 3H_2O$

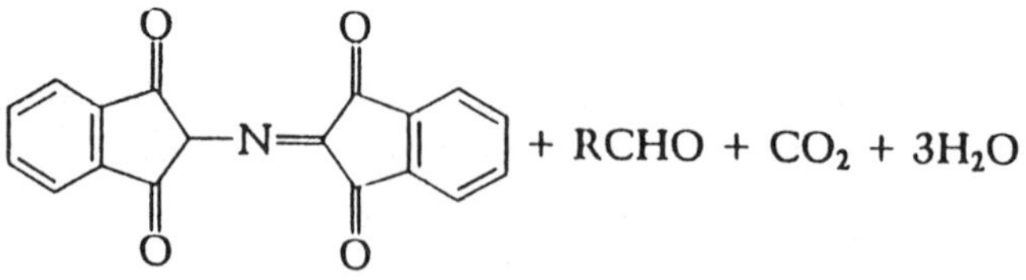

purple-blue complex

3. <u>Heavy Metal Ions Test.</u> Heavy metal ions precipitate proteins from their solutions. The ions that are most commonly used for protein precipitation are Zn^{2+}, Fe^{3+}, Cu^{2+}, Sb^{3+}, Ag^+, Cd^{2+}, and Pb^{2+}. Among

these metal ions, Hg^{2+}, Cd^{2+}, and Pb^{2+} are known for their notorious toxicity. They can cause serious damage to proteins (especially the enzymes) by denaturing them. This can result in death. The precipitation occurs because proteins become cross-linked by heavy metals as shown below:

$$2NH_2\text{---}\overset{\displaystyle O}{\overset{\|}{C}}\text{---}O^- + Hg^{2+} \rightarrow$$

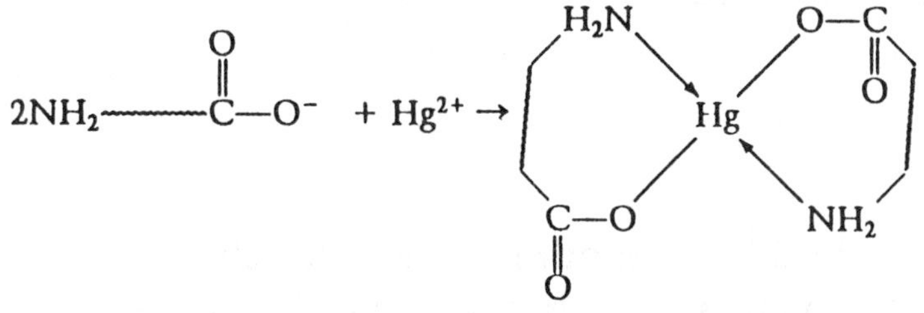

insoluble precipitate

Victims swallowing Hg^{2+} or Pb^{2+} ions are often treated with an antidote of a food rich in proteins which can combine with mercury or lead ions in the victim's stomach and, hopefully, prevent absorption! Milk and raw egg white are used most often. The insoluble complexes are then immediately removed from the stomach by an emetic.

4. The Xanthoprotein Test. This is a characteristic reaction of proteins that contain phenyl rings

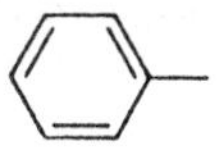

Concentrated nitric acid reacts with the phenyl ring to give a yellow-colored aromatic nitro compound. Addition of alkali at this point will deepen the color to orange.

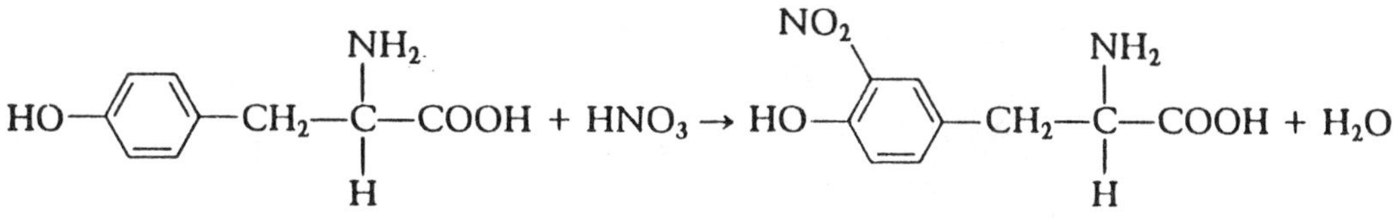

tyrosine colored compound

The yellow stains on the skin caused by nitric acid are the result of the xanthoprotein reaction.

Objectives

1. To isolate the casein from milk under isoelectric conditions.
2. To perform some chemical tests to identify proteins.

Procedure

A. Isolation of Casein

1. To a 250-mL Erlenmeyer flask, add 50.00 g of milk and heat the flask in a water bath (a 600-mL beaker containing about 200 mL of tap water; see Fig. 18.1). Stir the solution constantly with a stirring rod. When the bath temperature has reached about 40°C, remove the flask from the water bath and add about ten drops of glacial acetic acid while stirring. Observe the formation of a precipitate.

2. Filter the mixture into a 100-mL beaker by pouring it through a cheese cloth which is fastened with a rubber band over the mouth of the beaker (Fig. 18.2). Remove most of the water from the precipitate by squeezing the cloth gently. Discard the filtrate in the beaker. Using a spatula, scrape the precipitate from the cheese cloth into the empty flask.

3. Add 25 mL of 95% ethanol to the flask. After stirring the mixture for 5 min, allow the solid to settle. Carefully decant (pour off) the liquid that contains fats into a beaker. Discard the liquid.

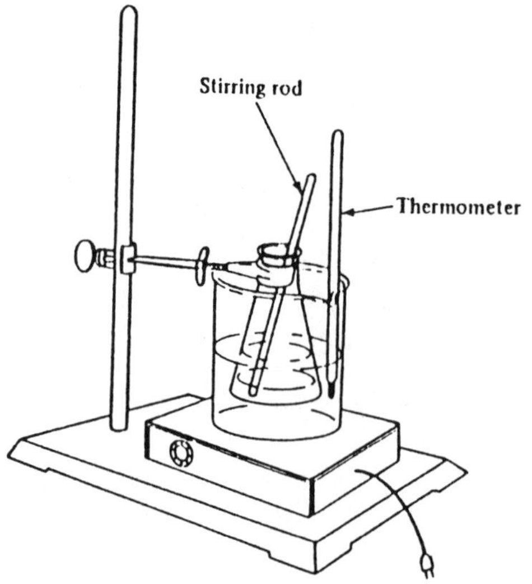

Figure 18.1 Precipitation of casein.

Figure 18.2 Filtration of casein.

4. To the residue, add 25 mL of 1:1 mixture of ether-ethanol. After stirring the resulting mixture for 5 min, collect the solid by vacuum filtration.

5. Spread the casein on a paper towel and let it dry. Weigh the dried casein and calculate the percentage of casein in the milk.

$$\% \text{ casein} = \frac{\text{weight of solid (casein)}}{\text{50.00 g of milk}} \times 100$$

B. Chemical Analysis of Proteins

1. Biuret Test. Place 15 drops of each of the following solutions in five clean, labeled test tubes.
   a. 2% glycine
   b. 2% gelatin
   c. 2% albumin
   d. Casein prepared in part A (one quarter of a full spatula) + 15 drops of distilled water
   e. 1% tyrosine

   To each of the test tubes, add five drops of 10% NaOH solution and two drops of a dilute $CuSO_4$ solution while swirling. The development of purplish violet color is evidence of the presence of proteins. Record your results on the Report Sheet.

2. The Ninhydrin Test. Place 15 drops of each of the following solutions in five clean, labeled test tubes.
   a. 2% glycine
   b. 2% gelatin
   c. 2% albumin
   d. Casein prepared in part A (one quarter of a full spatula) + 15 drops of distilled water
   e. 1% tyrosine

   To each of the test tubes, add five drops of ninhydrin reagent and heat the test tubes in a boiling water bath for about 5 min. Record your results on the Report Sheet.

3. Heavy Metal Ions Test. Place 2 mL of milk in each of three clean, labeled test tubes. Add a few drops of each of the following metal ions to the corresponding test tubes as indicated below:
   a. Pb^{2+} as $Pb(NO_3)_2$ in test tube no. 1
   b. Hg^{2+} as $Hg(NO_3)_2$ in test tube no. 2
   c. Na^+ as $NaNO_3$ in test tube no. 3

   Record your results on the Report Sheet.

CAUTION! **The following test will be performed by your instructor.**

4. The Xanthoprotein Test. (Perform the experiment under hood). Place 15 drops of each of the following solutions in five clean, labeled test tubes:
   a. 2% glycine
   b. 2% gelatin
   c. 2% albumin
   d. Casein prepared in part A (one quarter of a full spatula) + 15 drops of distilled water
   e. 1% tyrosine

   To each test tube, add ten drops of concentrated HNO_3 while swirling. Heat the test tubes carefully in a warm water bath. Observe any change in color. Record the results on your Report Sheet.

Chemicals and Equipment

1. Hot plate
2. Buchner funnel in no. 7 one-hole rubber stopper
3. 500-mL filter flask
4. Filter paper, (Whatman no.2) 7 cm
5. Cheese cloths
6. Rubber band
7. Boiling chips
8. 95% ethanol
9. Ether-ethanol mixture
10. Regular milk
11. Glacial acetic acid
12. Concentrated nitric acid
13. 2% albumin
14. 2% gelatin
15. 2% glycine
16. 5% copper(II) sulfate
17. 5% lead(II) nitrate
18. 5% mercury(II) nitrate
19. Ninhydrin reagent
20. 10% sodium hydroxide
21. 1% tyrosine
22. 5% sodium nitrate

NAME SECTION DATE

PARTNER GRADE

Experiment 18

PRE-LAB QUESTIONS

1. Casein has an isoelectric point at pH 4.6. What kind of charges will be on the casein in its native environment, in milk?

2. Draw the structure of a peptide bond in a typical protein molecule.

3. Biuret is the most common test for proteins. Would an amino acid, such as proline, give a positive biuret test?

4. What are the three most toxic heavy metal ions?

NAME ______________________ SECTION __________ DATE __________

PARTNER ______________________ GRADE __________

Experiment 18

REPORT SHEET

Isolation of casein

1. Weight of milk 50.00 g
2. Weight of dried casein ____g
3. Percentage of casein in milk ____%

Chemical analysis of proteins

Biuret test

Substance	Color formed
2% glycine	
2% gelatin	
2% albumin	
casein + H_2O	
1% tyrosine	

Which of these chemicals gives a positive test with this reagent?______________________

Ninhydrin test

Substance	Color formed after heating
2% glycine	
2% gelatin	
2% albumin	
casein + H_2O	
1% tyrosine	

Which of these chemicals gives a positive test with this reagent? ______________________

Heavy metal ion test

Substance	Precipitates formed
$Pb(NO_3)_2$	
$Hg(NO_3)_2$	
$NaNO_3$	

Which of these metal ions gives a positive test with casein in milk? ______________________

Xanthoprotein test

Substance	Color formed before or after heating
2% glycine	
2% gelatin	
2% albumin	
casein + H_2O	
1% tyrosine	

Which of these chemicals gives a positive test with this reagent? ______________________

POST-LAB QUESTIONS

1. Explain why casein precipitates when acetic acid is added to it.

2. In the isolation of casein following the acidification, you removed the precipitate by filtering through a cheesecloth and squeezing the cloth. If you did not squeeze out all the liquids would your yield of casein different? Explain.

3. What functional group(s) will give a positive reaction with the ninhydrin reagent?

4. If by mistake (don't try it) your finger touched nitric acid and you observe a yellow color on your fingers, what functional group(s) in your skin is (are) responsible for this reaction?

5. Why is milk or raw egg used as an antidote in cases of heavy metal ion poisoning?

6. Name another amino acid in proteins besides tyrosine which would also give a color in the xanthoprotein test.

Experiment 19

Isolation and identification of DNA from yeast.

Background

Hereditary traits are transmitted by genes. Genes are parts of giant deoxyribonucleic acid (DNA) molecules. In lower organisms, such as bacteria and yeast, both DNA and RNA (ribonucleic acid) occur in the cytoplasm while in higher organisms, most of the DNA is inside the nucleus and the RNA is outside the nucleus, in other organelles and in the cytoplasm.

In this experiment we will isolate DNA molecules from yeast cells. The first task is to break up the cells. This is achieved by a combination of different techniques and agents. Grinding up the cells with sand disrupts them and the cytoplasm of many yeast cells is spilled out. However, this is not a complete process. The addition of a detergent, hexadecyltrimethylammonium bromide, CTAB, accomplishes two functions: 1) it helps to solubilize cell membranes and and thereby further weakens the cell structure and 2) it helps to inactivate the nucleic acid degrading enzymes, nucleases, that are present. The addition of a chelating agent, ethylenediamine tetraacetate, EDTA, also inactivates these enzymes. EDTA removes the di and tri-valent cations which are necessary for the activity of nucleases. Without this inhibition the nucleases would degrade the nucleic acids to their constituent nucleotides. The final assault on the yeast cell is the osmotic shock. This is provided by a hypotonic saline-EDTA solution. The already weakened cells (by grinding and treatment with CTAB) will burst in the hypotonic medium and spill their contents, nucleic acids among them.

Once the nucleic acids are in solution they must be separated from the other constituents of the cell. First the protein molecules must be removed. Many of the proteins of the cell are strongly associated with nucleic acids. The addition of sodium perchlorate ($NaClO_4$) dissociates the proteins from nucleic acids. When the mixture is shaken with the organic solvent (chloroform- isoamyl alcohol) the proteins are denatured and they precipitate at the interface. At the same time the lipid components of the cells are dissolved in the organic solvent. Thus the aqueous layer will contain nucleic acids, small water soluble molecules and even some proteins as contaminants.

The addition of ethanol precipitates the large molecules (DNA, RNA and proteins) and leaves the small molecules in solution. DNA, being the largest fibrous molecule, forms thread-like precipitates that can be spooled off on a rod. The protein and RNA form a gelatinous precipitate that cannot be picked up by winding them on a glass rod. Thus the spooling separates DNA from RNA.

After the isolation of DNA we will probe its identity by the diphenylamine test. The blue color of this test is specific for deoxyribose and the appearance of blue color can be used to identify the deoxyribose containing DNA molecule.

<u>Flow diagram of the DNA isolation process.</u>

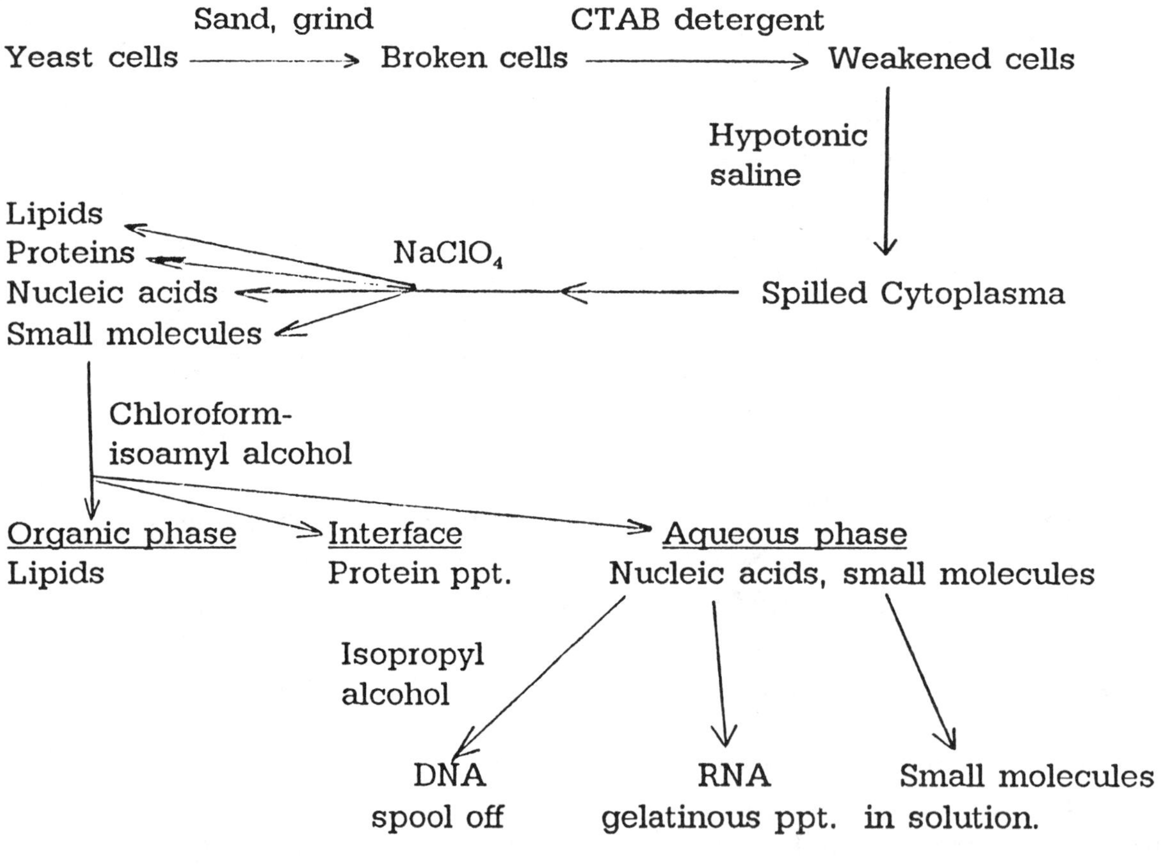

Objectives

To demonstrate the separation of DNA molecules from other cell constituents and to prove their identity.

Procedure

1. Cool a mortar in ice water. Add 2 to 3 g of baker's yeast and twice as much acid washed sand. Grind the yeast and the sand vigorously with a pestle for 5-10 minutes to disrupt the cells. (Two groups can work together in grinding and then divide the product.)

2. Preheat 25 mL of CTAB isolation buffer (2% CTAB, 0.15 M NaCl, 0.2% 2-mercaptoethanol, 20 mM EDTA and 100 mM Tris-HCl at pH 8.0) in a 100 mL beaker in a 60°C water bath.

3. Add the ground yeast and sand to the saline-CTAB solution. Mix the solution with the sand. Let it stand for 20 minutes, with occasional swirling, all the while maintaining the temperature at 60°C.

4. Decant the cell suspension into a 250 mL Erlenmayer flask leaving the sand behind. Cool the solution to room temperature. Add 5 mL of 6 M $NaClO_4$ solution and mix well. Transfer 40 mL chloroform-isoamyl alcohol mixture into the flask. Stopper the flask with a cork. Shake it for 10 minutes, sloshing the contents from side to side once every 15 second. A frothy emulsion will form. After 10 minutes let the emulsion settle.

5. Break up the emulsion by gently swirling with a glass rod reaching into the interface. The complete separation into two distinct layers is not possible without centrifugation. (If desk top centrifuges are available it is preferable to separate the layers by centrifuging at 1600 x g for 5 minutes). However, one can proceed without centrifugation as well. When sufficient amount (20-30 mL) of the top aqueous layer is cleared remove this with a Pasteur pipet and transfer it to a graduated cylinder. Measure the volume and pour the contents into a 250 mL beaker. Pay attention that none of the brownish precipitate, droplets of emulsion, is transferred.

6. To the viscous DNA containing aqueous solution, add slowly, twice its volume of cold isopropyl alcohol, taking care that the alcohol flows along the side of the beaker settling on top of the aqueous solution. With a flame sterilized glass rod gently stir the DNA-isopropyl alcohol solution. This procedure is <u>critical.</u> The DNA will form a thread-like precipitate. <u>Rotating (not stirring) the glass rod</u> spools all the DNA precipitate onto the glass rod. As the DNA is wound on the rod squeeze out the excess liquid by pressing the rod against the wall of the beaker. Transfer the spooled DNA on

the rod into a test tube containing 95 % ethanol.

7. Discard the alcohol solution left in the beaker and the chloroform-isoamyl alcohol solution left in the Erlenmayer flask into specially labelled waste jars. Do not pour them down the sink.

8. Remove the rod and the spooled DNA from the test tube. Dry the DNA with a clean filter paper. Note its appearance. Dissolve the isolated crude DNA in 2 mL citrate buffer (0.15 M NaCl, 0.015 M sodium citrate). Set up four dry and clean test tubes. Into the test tubes add 2 mL each of the following:

Test tube	Solution
1	1% glucose
2	1% ribose
3	1% deoxyribose
4	crude DNA solution

Add 5 mL diphenylamine reagent to each test tube.

<u>CAUTION:Diphenylamine reagent contains glacial acetic acid and concentrated sulfuric acid. Handle with care.</u>

Mix the contents of the test tubes. Heat the test tubes in boiling water bath for 10 minutes. Note the color.

Chemicals and Equipment

1. Baker's yeast
2. Sand
3. Saline-CTAB isolation buffer
4. $NaClO_4$ solution
5. Chloroform-isoamyl alcohol solvent
6. Citrate buffer
7. Isopropyl alcohol
8. Glucose solution
9. Ribose solution
10. Deoxyribose solution
11. Diphenylamine reagent
12. 95% ethanol
13. Mortar and pestle.
14. Desk top clinical centrifuges (optional)

NAME SECTION NAME

PARTNER GRADE

Experiment 19

PRE-LAB QUESTIONS

Consult Chapter 16 of your text book to answer the structural questions.

1. Draw the structure of the purine and pyrimidine bases that are part of the DNA molecule.

2. Draw the structures of ribose and deoxyribose.

3. Why must you handle the diphenylamine reagent with great care?

4. The most demanding part of this experiment is grinding the yeast cells with sand. What does this process accomplish? Would you be able to isolate DNA without this grinding?

NAME ____________________ SECTION __________ DATE __________

PARTNER ____________________ GRADE __________

Experiment 19

REPORT SHEET

1. Describe the appearance of the crude DNA preparation.

2. Diphenylamine test.

Solution	Color
1% glucose	____________
1% ribose	____________
1% deoxyribose	____________
crude DNA sample	____________

Did the diphenylamine test confirm the identity of DNA?

3. How did you separate the DNA from proteins?

4. After mixing the aqueous extract with chloroform-isoamyl alcohol mixture, which layer contained the RNA (aqueous or organic)?

5. What compounds were left behind in the isopropyl alcohol solution after spooling the DNA?

POST-LAB QUESTIONS

1. Can the diphenylamine reagent distinguish between ribose and deoxyribose and between DNA and RNA?

2. If you forgot to add $NaClO_4$ solution to your extract how would the omission effect the purity of your DNA preparation?

3. Why can we isolate DNA from the precipitate which also contains RNA and proteins by the simple "spooling" procedure?

EXPERIMENT 20

Kinetics of urease catalyzed decomposition of urea

Background

Enzymes speed up the rates of reactions by forming an enzyme-substrate complex. The reactants can undergo the reaction on the surface of the enzyme, rather than finding each other by collision. Thus, the enzyme lowers the energy of activation of the reaction. In this experiment we investigate the rate of the urea decomposition reaction:

$$H_2N-\overset{\displaystyle O}{\overset{\|}{C}}-NH_2 + H_2O \rightleftarrows CO_2 + 2NH_3 \qquad (1)$$

This reaction is catalyzed by a highly specific enzyme, urease. Urease is present in a number of bacteria and plants. The most common source of the enzyme is jack bean or soy bean. Urease was the first enzyme that was crystallized. Sumner, in 1926, proved unequivocally that enzymes are protein molecules.

Urease is a -SH group (thiol) containing enzyme. The cysteine residues of the protein molecule must be in the reduced -SH form in order for the enzyme to be active. Oxidation of these groups will form -S-S-, disulfide bridges and the enzyme loses its activity. Reducing agents such as cysteine or glutathione can reactivate the enzyme.

Heavy metals such as Ag^+, Hg^{2+}, or Pb^{2+}, which form complexes with the -SH groups, also inactivate the enzyme. For example, the poison phenylmercuric acetate is a potent inhibitor of urease.

$$\underset{\text{active}}{\text{enzyme}-SH} + \underset{\text{phenylmercuric acetate}}{CH_3-\overset{\displaystyle O}{\overset{\|}{C}}-O-Hg-C_6H_5} \rightleftarrows \underset{\text{acetic acid}}{CH_3\overset{\displaystyle O}{\overset{\|}{C}}-OH} + \underset{\text{inactive}}{\text{enzyme}-S-Hg-C_6H_5} \qquad (2)$$

In this experiment we study the kinetics of the urea decomposition. As shown in equation (1), the products of the reaction are carbon dioxide, CO_2, and ammonia, NH_3. Ammonia, being a base, can be titrated with an acid, HCl, and in this way we can determine the amount of NH_3 that is produced.

$NH_3 + HCl = NH_4Cl$ (3)

For example, a 5 mL aliquot of the reaction mixture is taken before the reaction starts. We use this as a blank. We titrate this with 0.05 N HCl to an end point. The amount of acid used was 1.5 mL. This blank then must be subtracted from all subsequent titration values. Next, we take a 5-mL sample of the reaction mixture after the reaction had proceeded for 10 min. We titrate this with 0.05 N HCl and, let's assume, get a value of 5.0 mL HCl. Therefore, 5.0 - 1.5 = 3.5 mL of 0.05 N HCl was used to neutralize the NH_3 produced in a 10 min. reaction time. This means that

(3.5 ~~mL~~ x 0.05 moles HCl)/1000 ~~mL~~ = 1.75 x 10^{-4} moles HCl was used up.

According to reaction (3), one mole of HCl neutralizes 1 mole of NH_3, therefore, the titration indicates that in our 5-mL sample 1.75 x 10^{-4} moles of NH_3 was produced in 10 min. Equation (1) also shows that for each mole of urea decomposed, 2 moles of NH_3 are formed. Therefore, in 10 minutes

(1 mole urea x 1.75 x 10^{-4} ~~moles NH_3~~)/2 ~~moles NH_3~~ = 0.87 x 10^{-4} moles urea or
8.7 x 10^{-5} moles of urea

were decomposed. Thus, the rate was 8.7 x 10^{-6} moles of urea per minute. This is the result we obtained using a 5-mL sample in which 1 mg of urease was dissolved. This rate of reaction corresponds to 8.7 x 10^{-6} moles urea/mg enzyme-minute.

A unit of activity of urease is defined as the micromoles (1 x 10^{-6} moles) of urea decomposed in 1 min. Thus, the enzyme in the preceding example had an activity of 8.7 units per mg enzyme.

In this experiment we also study the rate of the urease catalyzed reaction in the presence of an inhibitor. We use a dilute solution of phenylmercuric acetate to inhibit but not completely inactivate urease.

CAUTION! Mercury compounds are poisons. Take extra care to avoid getting mercuric salt solution in your mouth or swallowing it.

Many of the enzymes in our body are also -SH containing enzymes, and these will be inactivated if we ingest such compounds. As a result of mercury poisoning, many body functions will be inhibited.

Objectives

1. To demonstrate how to measure the rate of an enzyme catalyzed reaction.
2. To investigate the effect of an inhibitor on the rate of reaction.
3. To calculate urease activity.

Procedure

Enzyme kinetics in the absence of inhibitor.

1. Prepare a 37°C water bath in a 250-mL beaker. Maintain this temperature by occasionally adding hot water to the bath. To a 100-mL Erlenmeyer flask add 20 mL of 0.05 M Tris buffer and 20 mL of 0.3 M urea in a Tris buffer. Mix the two solutions and place the corked Erlenmeyer flask into the water bath for 5 min. This is your **reaction vessel.**

2. Set up a buret filled with 0.05 N HCl. Place into a 100-mL Erlenmeyer flask three to four drops of a 1% $HgCl_2$ solution. This will serve to stop the reaction, once the sample is pipetted into the titration flask. Add a few drops of methyl red indicator. This Erlenmeyer flask will be referred to as the **titration vessel.**

3. Take out the **reaction vessel** from the water bath. Add 10 mL of urease solution to your reaction vessel. The urease solution contains a specified amount of enzyme (e.g., 20 mg enzyme in 10 mL of solution). Note the time of adding the enzyme solution as zero reaction time. Immediately pipet a 5 mL aliquot of the urea mixture into your **titration vessel**. Stopper the reaction vessel and put it back into the 37°C bath.

4. Titrate the contents of the titration vessel with 0.05 N HCl to an end point. The end point is reached when the color changes from yellow to pink and stays that way for 10 seconds. Record the amount of acid used. This is your blank.

5. Wash and rinse your titration vessel after each titration and reuse it for subsequent titrations.

6. Take a 5-mL aliquot from the reaction vessel every 10 minutes. Pipet these aliquots into the cleaned titration vessel into which methyl red indicator and $HgCl_2$ inhibitor were already placed similarly to the procedures you used in your first titration (blank). Record the time you placed the aliquots into the titration vessels and titrate them with HCl to an end point. Record the amount of HCl used in your titration. Use five samples over a period of 50 min.

Enzyme kinetics in the presence of inhibitor

CAUTION! Be careful with the phenylmercuric acetate solution. Do not get it in your mouth or eyes.

1. Use the same water bath as in the first experiment. Maintain the temperature at 37°C. To a new 100-mL reaction vessel, add 19 mL of 0.05 M Tris buffer, 20 mL of 0.3 M urea solution, and 1 mL of phenylmercuric acetate (1×10^{-3} M). Mix the contents and place the reaction vessel into the water bath for 5 min.
2. Ready the titration vessel as before by adding a few drops of $HgCl_2$ and methyl red indicator. To the reaction vessel, add 10 mL of urease solution. Note the time of addition as zero reaction time. Mix the contents of the reaction vessel. Transfer immediately a 5-mL aliquot into the titration vessel. This will serve as your blank.
3. Titrate it as before. Record the result. Every 10 min. take a 5 mL aliquot for titration. The duration of this experiment should be 40 min.

Chemicals and Equipment

1. Tris buffer
2. 0.3 M urea
3. 0.05 N HCl
4. 1×10^{-3} M phenylmercuric acetate
5. 1% $HgCl_2$
6. Methyl red indicator
7. Urease solution

8. 50-mL buret
9. 10-mL graduated pipets
10. 5-mL volumetric pipets
11. 10-mL volumetric pipets
12. Buret holder
13. Spectroline pipet filler

NAME SECTION DATE

PARTNER GRADE

Experiment 20

PRE-LAB QUESTIONS

1. To what enzyme group does urease belong?

2. What groups are in the active site of urease?

3. Why is phenylmercuric acetate such a dangerous poison?

4. How do we measure the concentration of a product from the urease catalyzed reaction?

NAME ____________ SECTION ______ DATE ______

PARTNER ____________ GRADE ______

Experiment 20

REPORT SHEET

Enzyme kinetics in the absence of inhibitor

Reaction time (min)	Buret readings before titration (A)	Buret readings after titration (B)	mL acid titrated (B)-(A)	mL 0. 05 N HCl used up in the reaction (B)-(A)-blank
0 (blank)				
10				
20				
30				
40				
50				

Enzyme kinetics in the presence of Hg salt inhibitor

Reaction time (min)	Buret readings before titration (A)	Buret readings after titration (B)	mL acid titrated (B)-(A)	mL 0.05 N HCl used up in the reaction (B)-(A)-blank
0 (blank)				
10				
20				
30				
40				

NAME ______________________ SECTION __________ DATE ________

PARTNER ______________________ GRADE ______________

Experiment 20

REPORT SHEET

1. Present the preceding data in the graphical form by plotting column 1 on the x-axis and column 5 on the y-axis for both reactions.

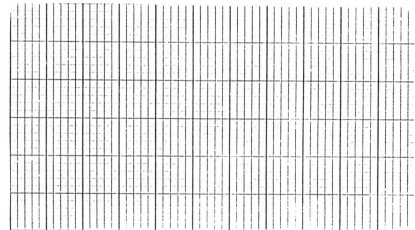

2. Calculate the urease activity only for the reaction without the inhibitor. Use the titration data from the first 10 min of reaction (initial slope).

Urease activity =

$$= \frac{\text{X } \cancel{\text{mL HCl}} \text{ consumed} \times 0.05 \ \cancel{\text{moles HCl}} \times 1 \ \cancel{\text{mole } NH_3} \times 1 \text{ mole urea} \times 50 \ \cancel{\text{mL sol}}}{10 \text{ min} \times 5 \ \cancel{\text{mL sol}} \times 1000 \ \cancel{\text{mL HCl}} \times 1 \ \cancel{\text{mole HCl}} \times 2 \ \cancel{\text{moles } NH_3} \times 20 \text{ mg urease}}$$

= Z units activity/mg enzyme

POST-LAB QUESTIONS

1. What would be the urease activity if you used the slope between 40 and 50 min. instead of the initial slope from your diagram?

2. Your instructor will provide the activity of urease as it was specified by the manufacturer. Compare this activity with the one you calculated. Can you account for the difference? (Enzymes usually lose their activity in long storage).

3. What was the purpose of adding $HgCl_2$ to the titration vessel?

4. If you did not add a few drops of $HgCl_2$ to the titration vessel, would the titration values of HCl used be higher or lower than the one you obtained?

5. In studying enzyme reactions, you must work at constant temperature and pH. What steps were taken in your experiment to satisfy these requirements?

6. Your lab ran out of 0.05 N HCl. You found a bottle labelled 0.05 N H_2SO_4. Could you use it for your titration? If you do, would your calculation of urease activity be different?

Experiment 21

Isocitrate dehydrogenase, an enzyme of the citric acid cycle

Background

The citric acid cycle is the first unit of the common metabolic pathway through which most of our food is oxidized to yield energy. In the citric acid cycle the partially fragmented food products are broken down further. The carbons of the C_2 fragments are oxidized to CO_2, released as such and expelled in the respiration. The hydrogens and the electrons of the C_2 fragments are transferred to the coenzyme, nicotineamide adenine dinucleotide, NAD^+, or to flavin adenine dinucleotide, FAD, which in turn become NADH + H^+ or $FADH_2$, respectively. These enter the second part of the common pathway, oxidative phosphorylation, and yield water and energy in the form of ATP.

The first enzyme of the citric acid cycle to catalyze both the release of one carbon dioxide and the reduction of NAD^+ is isocitrate dehydrogenase. The overall reaction of this step:

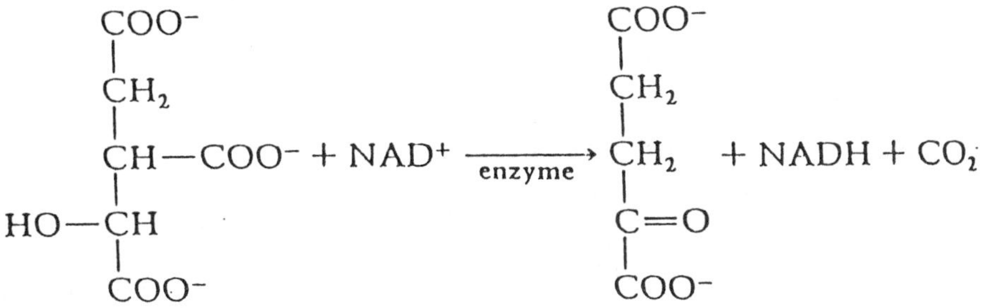
COO⁻
CH₂
CH—COO⁻ + NAD⁺
HO—CH
COO⁻
enzyme
COO⁻
CH₂
CH₂
C=O
COO⁻
+ NADH + CO₂

Isocitrate α-Ketoglutarate

The reduction of the NAD^+ itself is given by the equation:

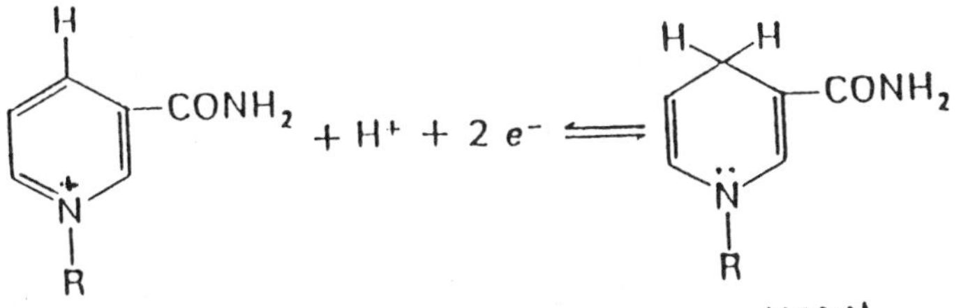
H
CONH₂
N
R
+ H⁺ + 2 e⁻
H H
CONH₂
N
R

NAD⁺ NADH

The enzyme has been isolated from many tissues, the best source being a heart muscle or yeast. The isocitrate dehydrogenase requires the presence of cofactors Mg^{2+} or Mn^{2+}. As an allosteric enzyme it is regulated by a number of modulators. ADP, adenosine diphosphate, is a positive modulator and therefore, stimulates enzyme activity. The enzyme has an optimum pH of 7.0. As is the case with all enzymes of the citric acid cycle, isocitrate dehydrogenase is found in the mitochondria. To isolate the enzyme, the cells of the yeast must be disintegrated which can be done by grinding them with sand.

In the present experiment you will determine the activity of isocitrate dehydrogenase extracted from baker's yeast. The basis of the measurement of the enzyme activity is the absorption spectrum of NADH. This reduced coenzyme has an absorption maximum at 340 nm. Therefore, an increase in the absorbance at 340 nm indicates an increase in NADH concentration, hence the progress of the reaction. We define the unit of isocitrate dehydrogenase activity as one that causes an increase of 0.01 absorbance per minute at 340 nm.

For example, if a 10 mL solution containing isocitrate and yeast extract exhibits a 0.04 change in the absorbance in two minutes, the enzyme activity will be

$$\frac{0.04\ \cancel{\text{abs.}}}{2\ \cancel{\text{min}} \times 10\ \text{mL}} \times \frac{1\ \text{unit}}{0.01\ \cancel{\text{abs.}}/1\ \cancel{\text{min}}} = 0.2\ \text{units/mL}$$

Objectives

To isolate an enzyme of the citric acid cycle, isocitrate dehydrogenase, and to measure its activity.

Procedure

A. <u>Enzyme extraction.</u> It is preferable to work in pairs. If not enough mortars and pestles are available four students can work together in preparing the enzyme extract. In the latter case, double the amounts recommended. Cool a mortar in ice water. Weigh 2 g of fresh baker's yeast and place it in the mortar. Record the weight of the yeast on the Report Sheet (1). Add approximatly 4 g of acid washed sand. Grind the yeast and sand with a pestle for 10 minutes to break up the cells. Add 5 mL ice cold 0.1 M $NaHCO_3$ solution. Record the volume on your Report Sheet (2). Continue grinding with the pestle vigorously for five minutes. Allow the sand and broken yeast fragments to settle out of the solution for 5 minutes.

B. Light absorption measurements. This part can also be performed in pairs to reduce the number of spectrophotometers needed. The detailed instructions how to operate the spectrophotometer depends on the make of the instrument and will be provided by your lab instructor. The general outlines of the operation are given below:

1. While waiting for the sand to settle in the mortar, turn on the instrument and let it warm up for a few minutes. Turn the wavelength control knob to read 340 nm. With no sample tube in the sample compartment, adjust amplifier control knob so that 0 % transmittance or infinite absorbance is read.
2. Fill a sample tube with distilled water and insert it into the spectrophotometer. Adjust the reading to 100 % transmittance (or 0 absorbance). This zeroing must be performed every 5-10 minutes since some instruments have the tendency to drift. The instrument is now ready to measure enzyme activity.
3. Decant the yeast extract from the mortar into a clean test tube labelled "extract". If your extract looks brown and turbid you must purify it by vacuum filtration (Fig. 21.1).

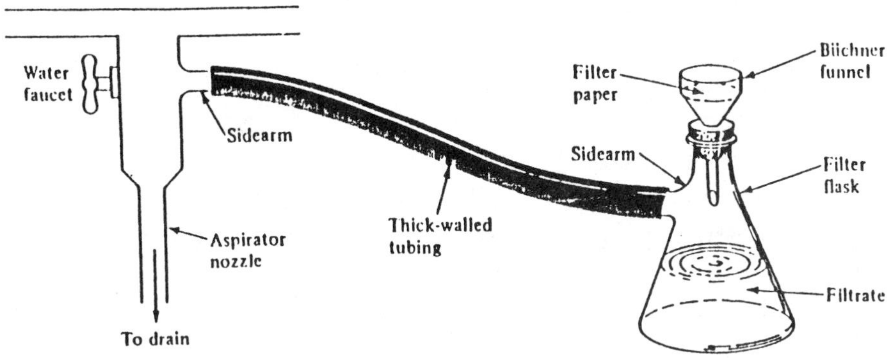

Figure 21.1 Vacuum filtration apparatus.

Place a Whatman No.1 filter paper in the Buchner funnel. Wet it with a few mL of distilled water and apply suction by opening the water faucet. The filter paper must fit snugly and adher to the Buchner funnel. Weigh about 1 g of Celite filter aid in a 50 mL beaker. Add 25 mL distilled water and make a slurry by stirring it. While the Buchner funnel is under suction pour the slurry on top of the filter paper. Let it filter until the Celite appears to be dry. While the water is still running disconnect the suction flask from the faucet by removing the rubber hose from the side arm of the suction flask. With this maneuver you prevent the tap water entering into the suction flask. Remove the Buchner funnel and empty the water from the suction flask. Rinse it with distilled water and empty it again. Insert the Buchner funnel back into the suction flask and reconnect it to the aspirator. While the Büchner funnel is under suction, pour your yeast extract on top of the Celite filter aid. The filtrate coming through should be clear or only slightly turbid, otherwise the entire filtration procedure must be repeated. When the Celite filter aid appears to be dry your filtration is completed. Disconnect the rubber hose from the side arm of the suction flask. Turn off the faucet. Transfer the filtered yeast extract into a clean and dry 10-mL graduate cylinder and measure its volume. Report it on the Report Sheet (3).

4. Prepare two test tubes for the enzyme activity measurements. Add 0.4 mL of phosphate buffer to each test tube. Next add to each test tube 0.2 mL $MgCl_2$, 0.5 mL ADP and 0.5 mL NAD^+ solutions. Mix the contents of the test tubes. Label one of the test tubes "Blank" and add to it 1.0 mL yeast extract and 2.2 mL distilled water. Read its transmission (absorbance) in the spectrophotometer at 340 nm, and record this value on the Report Sheet (6).
5. Label the second test tube "Sample". Add to it 1.0 mL yeast extract. Record this value on the Report Sheet (4). Add 1.2 mL isocitrate solution. Note the time of the isocitrate addition. This is zero time. Add also 1.0 mL distilled water and mix the contents. Record the total volume on the Report Sheet (5). Read the transmission (absorbance) of the "Sample" test tube 30 seconds after the addition of isocitrate and again after 60, 90 120, 150 and 180 seconds. Take a last reading 5 minutes after zero time. Record your observations.
6. If your enzyme activity was low and you did not get 0.1 - 0.5 absorbance increase during the 3 minute period, repeat the experiment as in step 5 using greater amount of enzyme extract. For example, if you use 2 mL enzyme extract in each test tube, you then should add only 1.2 mL distilled water to your "Blank" test tube and none to your "Sample" test tube.

Chemicals and Equipment

1. Baker's yeast
2. Sand
3. $NaHCO_3$ buffer
4. Phosphate buffer, pH 7.0
5. 2.5 mM ADP solution
6. 2.0 mM NAD^+ solution
7. 5.0 mM isocitrate solution
8. Celite filter aid
9. Büchner funnel
10. Filter flask
11. Filter paper

NAME ______ SECTION ______ DATE ______

PARTNER ______ GRADE ______

Experiment 21

PRE-LAB QUESTIONS

1. What reaction preceeds the isocitrate oxidation (dehydrogenation) in the citric acid cycle?

2. What reaction follows the isocitrate oxidation (dehydrogenation) in the citric acid cycle?

3. How do we measure NADH concentration?

4. What is the function of sand in the enzyme preparation?

NAME ______________________ SECTION __________ DATE __________

PARTNER ______________________ GRADE ______________________

Experiment 21

REPORT SHEET

1.Weight of baker's yeast ______________

2.Volume of $NaHCO_3$ added ______________

3.Volume of yeast extract after filtration ______________

4.Volume of the yeast extract in "Sample" ______________

5.Total volume of the "Sample" ______________

Net absorbance of sample= absorbance of sample-absorbance of blank

6.Absorbance of blank ______________

Time (s)	Absorbance of sample	Net absorbance of sample
30		
60		
90		
120		
150		
180		
400		

1. Plot your data: Net absorption versus time. At zero time the net absorption is zero.

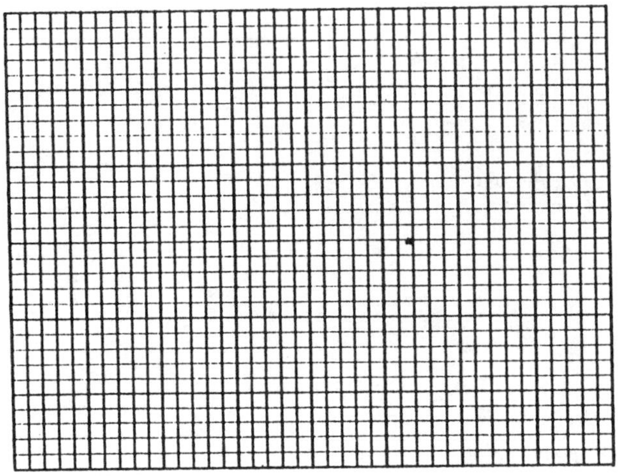

2. Calculate the enzyme activity:
   (a) Units of enzyme activity/ reaction mixture: The initial slope of the plot is usually a straight line. If so, read the value of net absorbance at 60 second. Divide it by 0.01.
   This gives you the number of enzyme activity units/reaction mixture.

   (One unit of enzyme activity is 0.01 net absorbance/minute.)

   (b) Calculate the isocitrate dehydrogenase activity per mg yeast.

Example: If the net absorbance/minute was 0.04. This is 4 units of activity. You used 0.2 mL of yeast extract from a total volume of 5 mL. This volume came from the original 2 g of baker's yeast.

$$\frac{4 \text{ units}}{0.2 \text{ mL extract}} \times \frac{5 \text{ mL extract}}{2 \text{ g yeast}} \times \frac{1 \text{ g yeast}}{1000 \text{ mg}} = 5 \times 10^{-2} \text{ units/ mg yeast}$$

POST-LAB QUESTION

1. What kind of enzyme activity would you observe if you did not grind the yeast sufficiently and only 10 % of the cells were broken?

2. You did your enzyme extraction at 4°C. Why was that necessary? Would you expect the enzyme activity to be the same or different if you would have done the extraction at room temperature?

3. Inspect your plot. You took the absorbance at 60 seconds in order to calculate the enzyme activity. Why did't you take the absorbance at 5 minutes? If you would have used the 5 minute value would your enzyme activity be higher, lower or the same as the one you calculated? Explain.

EXPERIMENT 22
Quantitative analysis of vitamin C contained in foods

Background

Vitamin C, also known as ascorbic acid, was one of the first vitamins which played a role in establishing the relationship between a disease and its prevention by proper diet. The disease, scurvy, has been known for ages and a vivid description of it was given by Jacques Cartier, a 16th century explorer of the American continent:"....Some did lose their strength and could not stand on their feet......Others.... had their skin spotted with spots of blood.....their mouth became stinking, their gums so rotten that all the flesh did fall off...." Prevention of scurvy was prescribed by eating fresh vegetables and fruits. The active ingredient in fruits and vegetables that helps to prevent scurvy is ascorbic acid which primates, including humans, cannot synthesize. Ascorbic acid is an important biological antioxidant (reducing agent). It helps to keep the iron in the enzyme, prolyl hydroxylase, in the reduced form and thereby it helps to maintain the activity of the enzyme. Prolyl oxidase, on the other hand, is essential for the synthesis of normal collagen. In scurvy abnormal collagen is synthesized that causes skin leisons and broken blood vessels.

The minimum daily requirement of vitamin C to prevent scurvy is 30 mg. Some people, among them most prominently, Linus Pauling, twice Nobel Laureate, advocate megadoses of vitamin C (250-10,000 mg/day) to prevent cancer, common cold etc.

In this experiment we will determine the vitamin C content of certain foods by titrating the solution with a water soluble form of iodine, I_3^-:

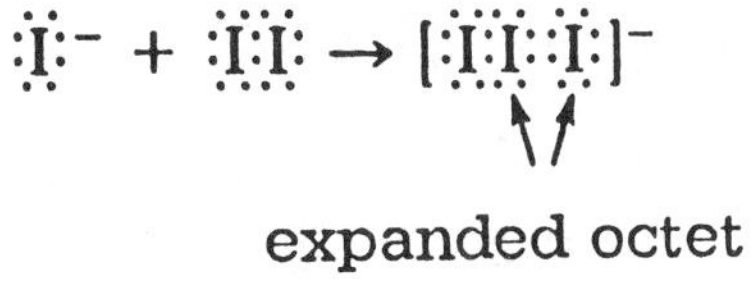

(tri-iodide ion)

Vitamin C is oxidized by I_2 (as I_3^-) according to the following chemical reaction:

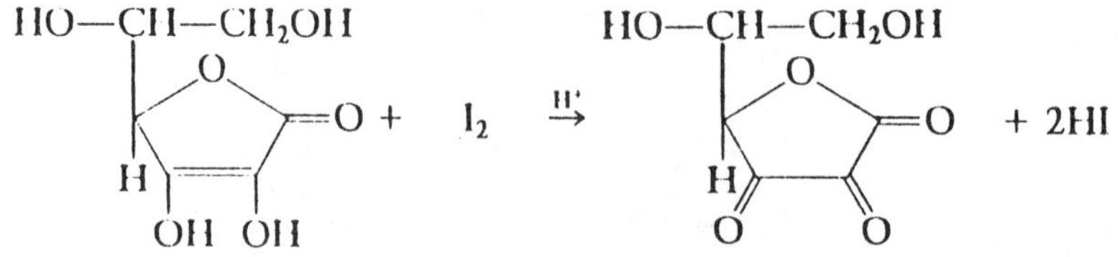

vitamin C (MW 176) | (MW 254) | oxidized product (MW 174)

As vitamin C is oxidized by iodine, I_2 becomes reduced to I^-. When the end point is reached (no vitamin C is left), the excess of I_2 will react with a starch indicator to form a starch-iodine complex which is blackish-blue in color.

I_2 + starch   iodine - starch complex (blackish-blue)

It is worthwhile to know that although vitamin C is very stable when dry, it is readily oxidized by air (oxygen) when in solution; therefore, a solution of vitamin C should not be exposed to air for long. The amount of vitamin C can be calculated by using the following conversion factor:

1 mL of I_2 (0.01 M) = 1.76 mg vitamin C

Objective

To determine the amount of vitamin C that is present in certain commercial food products by the titration method.

Procedure

1. Pour about 60 mL of a fruit drink that you wish to analyze into a clean, dry 100-mL beaker. Record the kind of drink on the Report Sheet (1).

2. If the fruit drink is cloudy or contains suspended particles, it can be clarified by the following procedure: Celite, used as a filter aid (about 0.5 g) is added to the fruit drink. After swirling it thoroughly, filter the solution through a glass funnel, bedded with a large piece of cotton. Collect the filtrate in a 50-mL Erlenmeyer flask (Fig. 22.1).

3. Using a 10-mL volumetric pipet and a Spectroline pipet filler, transfer 10.00 mL of the fruit drink into a 125-mL Erlenmeyer flask. Then add 20 mL of distilled water, five drops of 3 M HCl (as a catalyst), and ten drops of 2% starch solution to the flask.

4. Clamp a clean, dry 50-mL buret onto the buret stand. Rinse the buret twice with 5-mL portions of iodine solution. Let the rinsings run through the tip of the buret and discard them. Fill the buret slightly above the zero mark with a standard iodine solution. Record the molarity of standard iodine solution (2). (A dry funnel may be used for easy transfer.) Air bubbles should be removed by turning the stopcock several times to force the air bubbles out of the tip. Record the initial reading of standard iodine solution to the nearest 0.02 mL (3a).

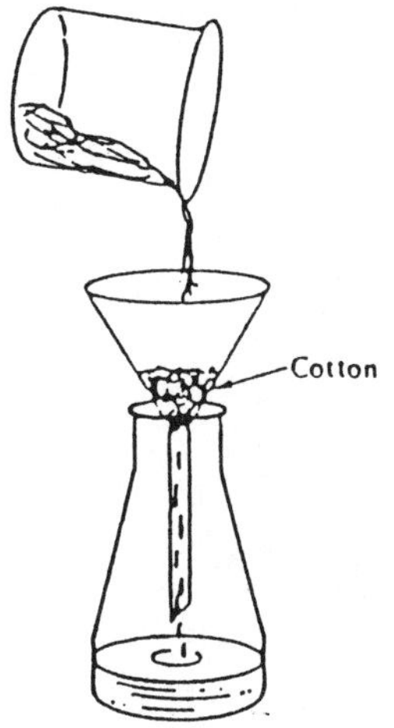

Figure 22.1 Clarification of fruit drinks.

5. Place the flask that contains the vitamin C sample under the buret and add the iodine solution dropwise to it while swirling, until the indicator just changes to dark blue. This color should persist for at least 20 seconds. Record the final buret reading (3b). Calculate the total volume of iodine solution that is required for the titration (3c), the weight of vitamin C in the sample (4), and % (w/v) of vitamin C in the drink (5). Repeat this titration procedure twice more, except using 20- and 30-mL portions of the same fruit drink instead of 10 mL. Record the volumes of iodine solution that are required for each titration.

Chemicals and Equipment

1. 50-mL buret
2. Buret clamp
3. Spectroline pipet filler
4. 10-mL volumetric pipet
5. Cotton
6. Filter aid
7. Hi-C apple drink
8. Hi-C orange drink
9. Hi-C grapefruit drink
10. 0.01 M iodine in potassium iodide
11. 3 M HCl
12. 2% starch solution

NAME ______ SECTION ______ DATE ______

PARTNER ______ GRADE ______

Experiment 22

PRE-LAB QUESTIONS

1. What are the symptoms of scurvy?

2. What natural foods contain appreciable amounts of vitamin C?

3. What is the minimum daily requirement of vitamin C to prevent scurvy in adults?

4. What are the megadoses of vitamin C advocated by Pauling?

NAME ____________ SECTION ________ DATE ________

PARTNER ____________ GRADE ____________

Experiment 22

REPORT SHEET

1. The kind of fruit drink ____________________

2. Molarity of iodine solution ____________________

3. Titration results

	Sample 1 (10.0 mL)	Sample 2 (20.0 mL)	Sample 3 (30.0 mL)
a. Initial buret reading	_____mL	_____mL	____mL
b. Final buret reading	_____mL	_____mL	____mL
c. Total volume of iodine solution used (b-a)	_____mL	_____mL	____mL
4. The weight of vitamin C in the fruit drink sample ((3C) x 1.76 mg/mL)	_____mg	_____mg	____mg
5. Concentration of vitamin C in the fruit drink ((4)/volume of drink)x100	_____mg/100mL	_____mg/100mL	____mg/100mL

6. Average concentration of vitamin C in the fruit drink _______ mg/100 mL

POST-LAB QUESTIONS

1. Why is HCl added for the titration of vitamin C?

2. What gives the blue color in your titration?

3. Write a chemical equation to show how vitamin C is oxidized by iodine.

EXPERIMENT 23

Urine Analysis

Background

The kidney is an important organ that filters materials from the blood that are harmful or in excess or both. These materials are excreted in the urine. A number of tests are routinely run in clinical laboratories on urine samples. These involve the measurements of glucose or reducing sugars, ketone bodies, albumin, specific gravity, and pH.

Normal urine contains little or no glucose, or reducing sugars; the amount varies from 0.05 to 0.15%. Higher concentrations may occur if the diet contains a large amount of carbohydrates or strenuous work was performed shortly before the test. Patients with diabetes or liver damage have chronically elevated glucose content in the urine. A semiquantitative test of glucose levels can be performed with the aid of test papers such as Clinistix. This is a quick test that uses a paper containing two enzymes: glucose oxidase and peroxidase. In the presence of glucose the glucose oxidase catalyzes the formation of gluconic acid and hydrogen peroxide. The hydrogen peroxide is decomposed with the aid of peroxidase and yields atomic oxygen.

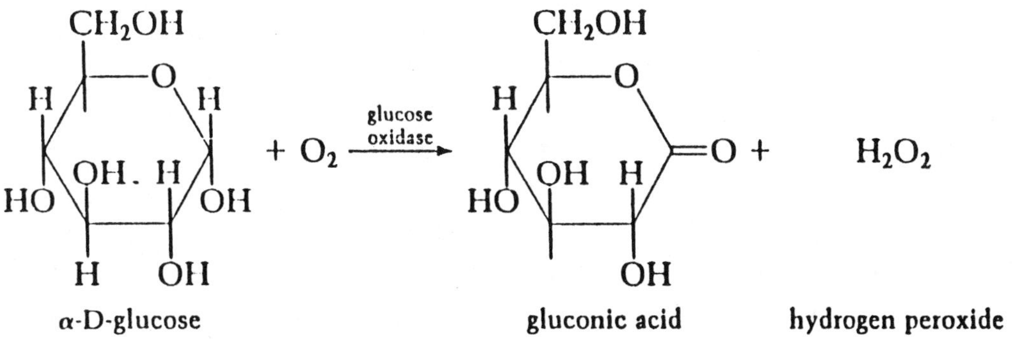
glucose
oxidase
α-D-glucose
gluconic acid
hydrogen peroxide

In Clinistix the atomic oxygen reacts with an indicator, o-tolidine, producing a purple color. The intensity of the purple color is proportional to the glucose concentration.

$$H_2O_2 \xrightarrow{\text{peroxidase}} H_2O + [O]$$

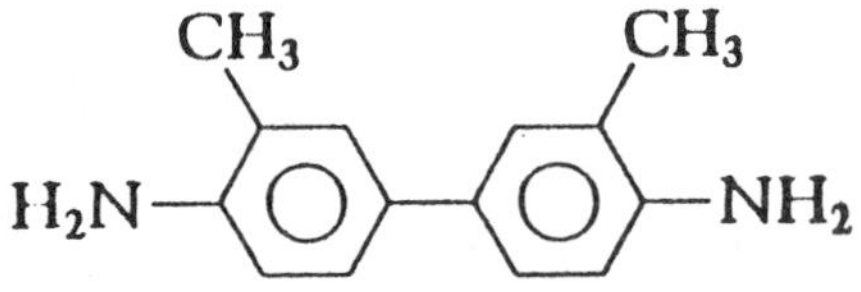

+ [O] → purple oxidation product

o-tolidine

This test is specific for glucose only. No other reducing sugar will give positive results.

Normal urine contains no albumin or only a trace amount of it. In case of kidney failure or malfunction the protein passes through the glomeruli and is not reabsorbed in the tubule. So, albumin and other proteins end up in the urine. The condition known as proteinuria may be symptomatic of kidney disease. The loss of albumin and other blood proteins will decrease the osmotic pressure of blood. This allows water to flow from the blood into the tissues, creating swelling (edema). Renal malfunction is usually accompanied by swelling of the tissues. The Albustix test is based on the fact that a certain indicator at a certain pH changes its color in the presence of proteins.

Albustix contains the indicator, tetrabromophenol blue, in a citrate buffer at pH 3. At this pH the indicator has a yellow color. In the presence of protein the color changes to green. The higher the protein concentration, the greener the indicator will be. Therefore, the color produced by the Albustix can be used to estimate the concentration of protein in urine.

Three substances that are the products of fatty acid catabolism - acetoacetic acid, β-hydroxybutyric acid, and acetone- are commonly called ketone bodies. These are normally present in the blood in small amounts and can be used as an energy source by the cells. Therefore, no ketone bodies will normally be found in the urine. However, when fats are the only energy source, excess production of ketone bodies will occur. They will be filtered out by the kidney and appear in the urine. Such abnormal conditions of high fat catabolism take place during starvation or in diabetes mellitus when glucose, although abundant, cannot pass through the cell membranes to be utilized inside where it is needed. Acetoacetic acid (CH_3COCH_2COOH), and to a lesser extent acetone (CH_3COCH_3) and β-hydroxybutyric acid ($CH_3CHOHCH_2COOH$), react with sodium nitroprusside ($Na_2(Fe(CN)_5NO))\cdot 2H_2O$ to give a maroon-colored complex. In Ketostix the test area contains a sodium nitroprusside and sodium phosphate buffer to provide the proper pH for the reaction. The addition of lactose to the mixture in the Ketostix enhances the development of the color.

Some infants are born with a genetic defect known as phenylketonuria (PKU). They lack the enzyme phenylalanine oxidase, which converts phenylalanine to tyrosine. Thus, phenylalanine accumulates in the body and it is degraded to phenylpyruvate by transamination;

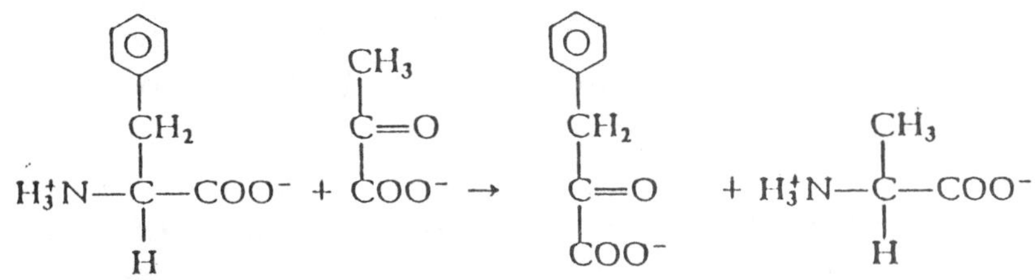

phenylalanine pyruvate phenylpyruvate alanine

Phenylpyruvate is excreted in the urine. Normal urine does not contain any phenylpyruvate. People suffering from PKU disease have varying amounts of phenylpyruvate in their urine. PKU disease causes severe mental retardation in infants if it is not treated immediately after birth by restricting the phenylalanine content of the diet. In many states the law requires that every newborn be tested for phenylpyruvate in the urine. The test is based on the reaction of the iron (III) ion with the phenylpyruvate, producing a gray-green color. Phenistix strips which are coated with $Fe(NH_4)(SO_4)_2$ and a buffer can detect as little as 8 mg of phenylpyruvate in 100 mL of urine. Some drugs such as aspirin produce metabolites (salicylic acid) that are excreted in the urine and give color with an iron (III) ion. However, this produces a deep purple color and not the gray-green of PKU. The purple color that is given by the Phenistix can be used to diagnose an overdose of aspirin. Other drugs, such as phenylthiazine in an overdose, give a gray-purple color with Phenistix. For PKU diagnosis only the appearance of the gray-green color means a positive test.

The creatinine content of normal urine is about 130 mg in 100 mL of urine. This is the anhydride derivative of creatine which is one of the storage materials of high energy phosphate, in the form of creatine phosphate, in the muscles. Creatinine is the degradation product of creatine and is excreted in the urine.

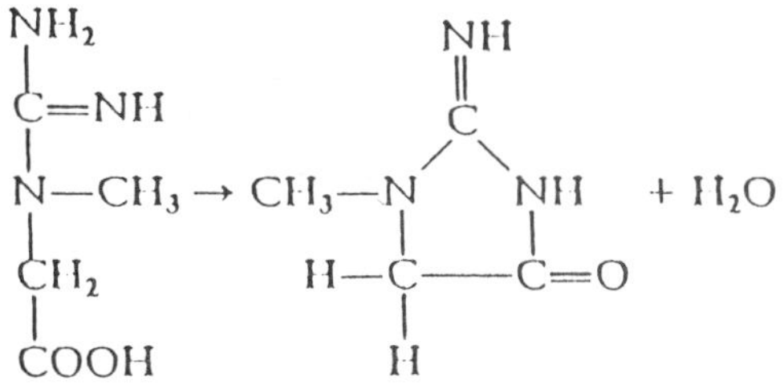

creatine creatinine

High creatinine values in urine analysis may indicate excessive breakdown of muscle creatine due to metabolic defects. In heart attacks the activity of creatine phosphatase increases. Consequently, the urine of heart attack patients may contain an abnormal amount of creatinine.

Urobilinogen and other bile pigments are normally minor components of urine (2 to 50 μg/100 mL). They are the products of hemoglobin breakdown. Bile pigments are usually excreted in the feces. In case of obstruction of the bile ducts (gallstones, obstructive jaundice), the normal excretion route through the small intestines is blocked and the excess bilirubin is filtered out of the blood by the kidneys and appears in the urine. Urobilistix is a test paper that can detect the presence of urobilinogen, because it is impregnated with para-dimethylaminobenzaldehyde. In strongly acidic media this reagent gives a yellow-brownish color with urobilinogen.

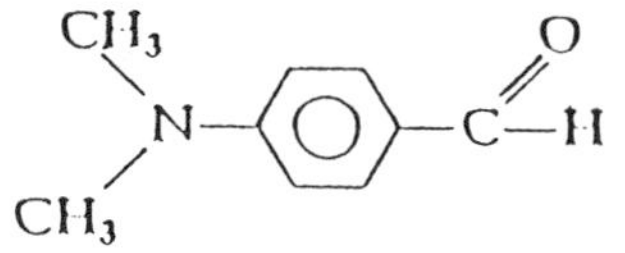

para-dimethylaminobenzaldehyde

The specific gravity of normal urine may range from 1.008 to 1.030. After excessive fluid intake (a beer party), the specific gravity may be on the low side; after heavy exercise and perspiration, it may be on the high side. High specific gravity indicates excessive dissolved solutes in the urine.

The pH of normal urine can vary between 4.7 to 8.0. The usual value is about 6.0. High protein diets and fever can lower the pH of urine. In severe acidosis the pH may be as low as 4.0. Vomiting and respiratory or metabolic alkalosis can raise the pH above 8.0.

Objectives

1. To perform quick routine analytical tests on urine samples.
2. To compare results obtained on "normal" and "pathological" urine samples.

Procedure

Each student must analyze her (his) own urine. A fresh urine sample will be collected prior to the laboratory period. The stockroom will provide paper cups for sample collection. While handling body fluids, such as urine, plastic gloves should be worn. The used body fluid will be collected in a special jar and disposed collectively. The plastic gloves worn will be collected and autoclaved before disposal.

In addition, the stockroom will provide one "normal" and two "pathological" urine samples.

Place 5 mL of urine sample from each source into four different test tubes. These will be tested with the different test papers.

Glucose Test

For the glucose test, use for comparison two test tubes, each half-filled, one with 0.25% and the other with 1% of a glucose solution. Take six strips of Clinistix from the bottle. Replace the cap immediately. Dip the test area of the Clinistix into one of the samples. Tap the edge of the strip against a clean, dry surface to remove the excess urine. Compare the test area of the strip to the color chart supplied on the bottle exactly 10 sec. after the wetting. Do not read the color changes that occur after 10 sec. Record your observation: the "light" on the color chart means 0.25% or less glucose; the "medium" means 0.4%; and the "dark" means 0.5% or more. Repeat the test with the other five samples.

Protein Test

For the protein test take four Albustix strips, one for each of the four urine samples, from the bottle. Replace the cap immediately. Dip the Albustix strips into the test solutions, making certain that the reagent area of the strip is completely immersed. Tap the edge of the strip against a clean, dry surface to remove the excess urine. Compare the color of the Albustix test area with the color chart supplied on the bottle. The time of the comparison is not critical, you can do it immediately or any time within a minute after the wetting. Read the color from yellow (negative) to different shades of green, indicating trace amounts of albumin up to 0.1%.

Ketone Bodies

To measure the ketone bodies' concentration in the urine, take four Ketostix strips from the bottle. Replace the cap immediately. Dip a Ketostix, one in each of the urine samples, and remove the strips at once. Tap them against a clean, dry surface to remove the excess urine. Compare the color of the test area of the strips to the color chart supplied on the bottle. Read the colors exactly 15 sec. after the wetting. A buff pink color indicates the absence of ketone bodies. A progression to a maroon color indicates increasing concentration of ketone bodies from 50 to 160 mg/L of urine.

Test for PKU

To test for PKU disease use four Phenistix strips, one for each urine sample. Immerse the test area of the strips into the urine samples and remove them immediately. Remove the excess urine by tapping the strips against a clean, dry surface. Read the color after 30 sec. of the wetting and match them against the color charts provided on the bottle. Record your estimated phenylpyruvate content: negative or 0.015 to 0.1%.

Urobilinogen

To measure the urobilinogen content of the urine samples, use Urobilistix strips, one for each urine sample. Dip them into the samples and remove the excess urine by tapping the strips against a clean, dry surface. Sixty sec. after the wetting, compare the color of the test area of the strips to the color chart provided on the bottle. Estimate the urobilinogen content and record it in Ehrlich units.

pH in Urine

To measure the pH of each urine sample, use pH indicator paper, such as pHydrion test paper within the pH range of 3.0 to 9.0.

For the preceding tests you may use a multipurpose strip such as Labstix that contains test areas for all these tests, except for the test for phenylpyruvate, on one strip. The individual test areas are separated from each other and clearly marked. The time requirements to read the colors are also indicated on the chart. The results of five tests can be read in 1 min.

Measure of Creatinine Content of Urine

To measure the creatinine content of the urine samples, place 2 mL of each urine sample into a clean, dry test tube. Take an additional three test tubes and add 2 mL of each of the three creatinine solutions: 0.1, 0.5, and 1%. Add eight drops of saturated picric acid solution to each test tube.

CAUTION! Picric acid is a strong acid. Avoid contact with it. If it gets on your skin, it will leave a yellow stain for a few days, even after washing it vigorously with soap. Solid picric acid is also an explosive. Take care not to drop the bottle with saturated solution of picric acid.

Add eight drops of 3 M NaOH solution to each test tube. Mix. Observe the color developed. Record your observation. Estimate the creatinine concentration of your urine samples by comparing the colors to those obtained on the creatinine solutions.

Specific Gravity of Urine

The specific gravity of your urine samples will be measured with the aid of a hydrometer (urinometer, see Fig. 23.1). Place the bulb in a cylinder. Add sufficient urine to the cylinder to make the bulb float. Read the specific gravity of the sample from the stem of the hydrometer where the meniscus of the urine intersects the calibration lines. Be sure the hydrometer is freely floating and does not touch the walls of the cylinder. In order to use as little urine as possible, the instructor may read the normal and two pathological urine samples for the whole class. If so, you, yourself, will measure the specific gravity of your own urine sample only.

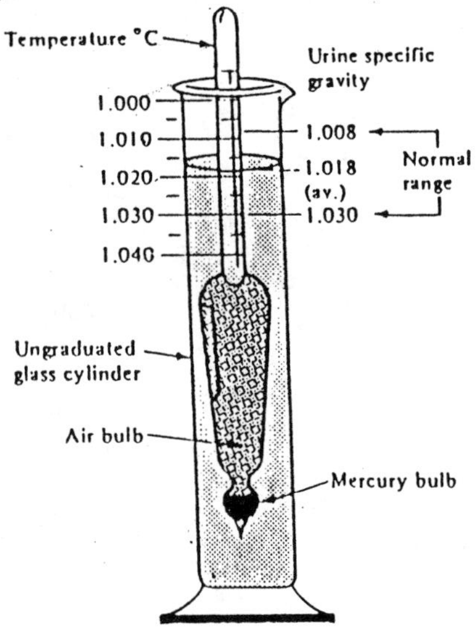

Figure 23.1 A urinometer.

Chemicals and Equipment

1. 0.25 and 1% glucose solutions
2. 0.1, 0.5, and 1% creatinine solutions
3. Clinistix
4. Albustix
5. Ketostix
6. Urobilistix
7. pH paper in the 3.0 to 9.0 range
8. A multipurpose Labstix (instead of these test papers)
9. Phenistix
10. Saturated picric acid solution
11. 3 M NaOH
12. Hydrometer (urinometer)

NAME SECTION DATE

PARTNER GRADE

Experiment 23

PRE-LAB QUESTIONS

1. Why should you avoid touching the picric acid solution?

2. In what tests do we use the following reagents?
   a. Fe^{3+}

   b. 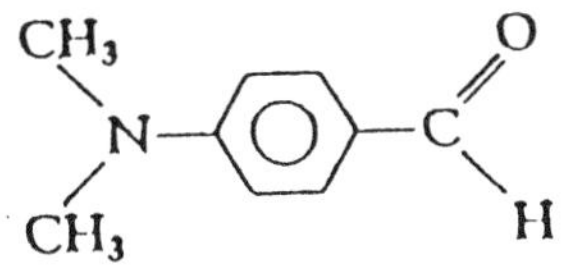

   c. 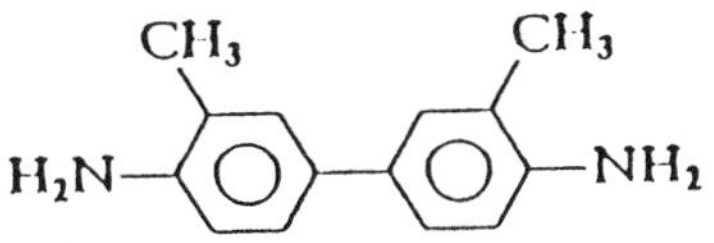

3. In determining glucose concentration in urine you use a Clinistix. Yet the purple color of a positive test is the result of the interaction of o-toluidine indicator with atomic oxygen. How is atomic oxygen related to glucose?

4. A patient's urine shows a high specific gravity, 1.04. The pH is 7.8, and the Phenistix test indicates a purple color that is not characteristic of PKU. The patient had a high fever for a few days and was given aspirin. Do these tests indicate any specific disease or are they symptomatic of recovering from a high fever? Explain.

NAME _______________ SECTION _______ DATE _______

PARTNER _______________ GRADE _______

Experiment 23

REPORT SHEET

Urine Samples

Test	Normal	Pathological A	Pathological B	Your own Remarks
Glucose				
Ketone bodies				
Albumin				
Urobilin-ogen				
pH				
Phenyl-pyruvate				
Creatinine				
Specific gravity				

POST-LAB QUESTIONS

1. Did you find any indication that your urine is not normal? If so, what may be the reason?

2. Why is the phenylpyruvate test mandatory with newborns in many states?

3. Urine samples came from three different persons. Each had a different diet for a week. One lived on a high protein diet; the second, on a high carbohydrate diet; and the third lived on a high fat (no carbohydrate) diet. How would their urine analyses differ from each other?

4. A patient's urine was tested with Clinistix and the color was read 60 s after wetting the strip. It showed 1.0% glucose in the urine. Is the patient diabetic? Explain.

EXPERIMENT 24

Measurement of sulfur dioxide preservative in foods

Background

Many foods contain preservatives that prolong the shelf life and/or combat infestations by insects and microorganisms. Sulfur dioxide is probably one of the oldest preservatives. For centuries people found that if the summer harvest of fruits is to be preserved and stored for the winter months, a drying process can accomplish the task. Raisins, dates, dried apricots and prunes are still sun-dried in many countries. The drying process increases the sugar concentration in such dried fruits and bacteria and most other microorganisms cannot use the dried fruit as a carbohydrate source because of the hypertonic (hyperosmotic) conditions.

It was found, by trial and error, that when storage areas are fumigated by burning sulfur, the dried fruits have a longer shelf life and are mostly void of insect and mold infestations as well. Sulfur dioxide, the product of the sulfur fumigation is still used today as a preservative. It is harmless when consumed in small quantities. The Food and Drug Administration requires the listing of sulfur dioxide on the labels of food products. You may see such listings on almost every bottle of wine, on packaged dried fruits and in some processed meat products.

In the present experiment we use a colorimetric technique to analyse the SO_2 content of raisins.

Objectives

1. To learn the use of standard curves for analysis.
2. To determine the SO_2 content by colorimetric analysis.

Procedure

A. Preparation of sample

1. Weigh 10 g of raisins and transfer it to a blender containing 290 mL distilled water. Record the weight of the raisins on your Report Sheet (1). This amount will be sufficient for a class of 25 students.

Cover and blend for 2 min.

2. Each student should prepare two 100-mL volumetric flasks. One will be labelled "blank", the other "sample". To each flask add 1 mL 0.5 N NaOH solution. To the "blank" add 10 mL distilled water. To the "sample" flask add 10 mL raisin extract. (Use a volumetric pipet to withdraw 10 mL from the bottom portion of the blender). Mix both solutions by swirling them for 15-30 sec.

3. Add to each volumetric flask 1 mL of 0.5 N H_2SO_4 solution and 20 mL of mercurate reagent. **Caution!** (Use polyethylene gloves to protect your skin from touching mercurate reagent. Mercurate reagent is toxic and if spills occur you should wash them immediatly with copious amounts of water). Add sufficient distilled water to bring both flasks to 100 mL volume.

B. Standard curve

1. Label five 100-mL volumetric flasks as No.1,2,3,4 and 5. To each flask add 5 mL mercurate reagent. Add standard sulfur dioxide solutions to the flasks as labeled (i.e. 1 mL to flask No.1, 2 mL to flask No.2 etc.). Bring each volumtric flask to 100 mL volume with distilled water.

2. Label five clean and dry test tubes as No.1,2,3,4 and 5. Transfer to each 5 mL portions of the corresponding samples (i.e. to test tube No.1 from volumetric flask No.1, etc.). Add 2.5 mL rosaniline reagent to each test tube. Add also 5 mL of 0.015% formaldehyde solution to each test tube. Cork the test tubes. Mix the content by shaking and swirling. Let it stand for 30 min. at room temperature. The SO_2 concentrations in your test tubes will be as follows:

Test tube No.	Concentration in ug/mL
1	10.0
2	20.0
3	30.0
4	40.0
5	50.0

Note these on your Report Sheet.

3. At the end of 30 min. read the intensity of the color in each test tube in a spectrophotometer. Your instructor will demonstrate the use of the spectrophotometer. (For reading absorbance values in spectrophotometers read the details in Experiment 41 under Procedure Part B.)

4. Construct your standard curve by plotting the absorbance readings on the y-axis and the corresponding concentration readings on the x-axis. Connect the points with the best straight line.

C. Measurement of SO_2 content of raisins.

1. Add 2.5 mL rosaniline reagent to each of four test tubes labeled "blank", No.1, No.2, and No.3. To these test tubes add the "sample" and "blank" you prepared in part A using the following scheme:

Test tube	"sample" mL	"blank" mL
Blank	0	2.5
No.1	0.5	2.0
No.2	1.0	1.5
No.3	2.0	0.5

2. To each test tube add 5 mL formaldehyde reagent. Mix by swirling and let it stand at room temperature for 30 min.

3. Read the absorbance of the solutions in your 4 test tubes and record it on your Report Sheet (2).

4. The "net" absorbance is the absorbance of the "sample" minus the absorbance of the "blank". Record the "net absorbance in (3). Using the standard curve obtained in Part B record the SO_2 content (in ug/mL) of your test tubes that correspond to your "net" absorbance values (4).

5. Calculate the SO_2 content of your raisin sample in each test tube. Record it on your Report Sheet (5). Here is a sample calculation for test tube No.2:

$$\frac{2.0\ SO_2\ \text{ug/mL solution from std.curve}}{1\text{mL sample}/\ 10\ \text{mL solution}} \times \frac{100\ \text{mL total sample}}{10\ \text{g dried fruit}} = 200\ \text{ug/g}$$

Average the three values obtained and record it on your Report Sheet (6).

Chemicals and Equipment

1. Raisins
2. Blender
3. 100-mL volumetric flasks
4. 10-mL pipet
5. 0.5 N NaOH solution
6. 0.5 N H_2SO_4 solution
7. 0.015% formaldehyde solution
8. Rosaniline reagent
9. Mercurate reagent
10. Sulfur dioxide stock solution
11. Spectrophotometers

NAME ______________________ SECTION __________ DATE __________

PARTNER ______________________ GRADE ______________________

Experiment 24

PRE-LAB QUESTIONS

1. What is a standard curve?

2. Write the balanced equation that shows that burning sulfur in air produces sulfur dioxide.

3. What is the difference between "blank" and "sample" solutions?

4. Your "standard sulfur dioxide stock solution" is actually sulfur dioxide gas dissolved in water. What is the product of this reaction? Write a balanced equation showing reactants and product.

NAME ____________________ SECTION __________ DATE __________

PARTNER ____________________ GRADE ____________________

Experiment 24

REPORT SHEET

Standard curve

Test tube #	Concentration of SO_2 in µg/mL	Absorbance
1	10.0	
2	20.0	
3	30.0	
4	40.0	
5	50.0	

Absorbance

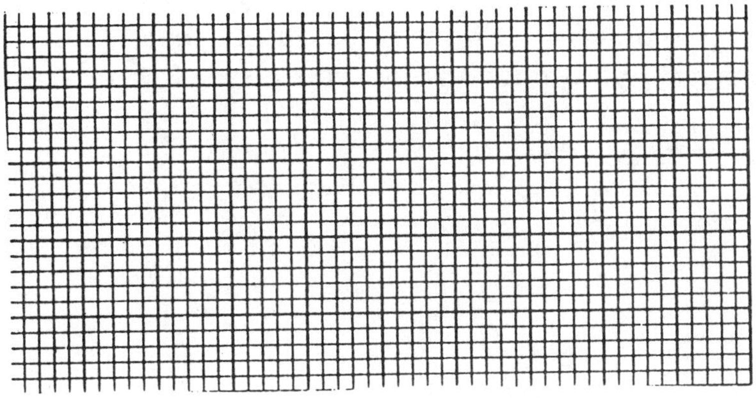

µg/mL

SO_2 content

Test tube	"sample" mL	Absorbance (2)	"Net" absorbance (3)	SO_2 µg/mL (4)	SO_2 µg/g (5)
Blank	0				
#1	0.5				
#2	1.0				
#3	2.0				

Average SO_2 content µg/g raisin ______________________(6)

POST-LAB QUESTIONS

1. Is your standard curve a straight line going through the origin?

2. If you made a mistake and calculated the SO_2 content using the absorbance values instead of the "net" absorbance values, would you get higher or lower than the true value of SO_2 concentration? Explain.

3. From the average value (6) of your SO_2 content, calculate how much <u>sulfur</u> do you ingest when you eat 100 g raisins.

EXPERIMENT 25

Measurement of the active ingredient in aspirin pills

Background

Medication delivered in the form of a pill contains an active ingredient or ingredients. Beside, the the drug itself, the pill also contains fillers. The task of the filler is many fold. Sometimes it is there to mask the bitter or otherwise unpleasant taste of the drug. Other times the filler is necessary because of the prescribed dose of the drug is so small in mass that it would be difficult to handle. Drugs that have the same generic name contain the same active ingredient. The dosage of the active ingredient must be listed as specified by law. On the other hand, neither the quantity of the filler nor its chemical nature appears on the label. That does not mean that the fillers are completely inactive. They usually effect the rate of drug delivery. In order to deliver the active ingredient the pill must fall apart in the stomach. For this reason many fillers are polysaccharides, for example, starch, that either are partially soluble in stomach acid or swell allowing the drug to be delivered in the stomach or in the intestines.

In the present experiment we measure the amount of the active ingredient, acetylsalicylic acid (see also Experiment 9) in common aspirin pills. Companies use different fillers and in different amounts, but the active ingredient, acetylsalicylic acid, must be the same in every aspirin tablet. We separate the acetylsalicylic acid from the filler based on their different solubilities. Acetylsalicylic acid is very soluble in ethanol while neither starch, nor other polysaccharides, or even mono- and disaccharides used as a fillers, are soluble in ethanol. Some companies may use inorganic salts as fillers, but these too are not soluble in ethanol. On the other hand, some special formulated aspirin tablets may contain small amounts of ethanol soluble substances such as stearic acid or vegetable oil. Thus, the ethanol extracts of aspirin tablets may contain small amounts of substances other than acetylsalicylic acid.

Objectives

1. To appreciate the ratio of filler to active ingredients in common aspirin tablets.
2. To learn techniques of quantitative separations.

Procedure

1. Weigh approximately 10 g aspirin tablets. Record the actual weight on your Report Sheet (1). Count the number of tablets and record it on your Report Sheet (2).

2. Place the weighed aspirin tablets in a mortar of approximately 100 mL capacity. Before starting to grind, place the mortar on a white sheet of paper and loosely cover it with a filter paper. The purpose of this procedure is to catch small fragments of the tablets that may fly out of the mortar during the grinding process. Break up the aspirin tablets by gently hammering them with the pestle. Recover and place back in the mortar any fragments that flew out during the hammering. With a twisting motion of your wrist grind the aspirin pieces into a fine powder with the aid of the pestle.

3. Add 10 mL of 95 % ethanol to the mortar and continue to grind for 2 minutes. Place a filter paper (Whatmann No.2, 7 cm) in a funnel and place the funnell in a 250-mL Erlenmeyer flask. With the aid of a glass rod transfer the supernatant liquid from the mortar to the filter paper. After a few minutes when about 1 mL of clear filtrate has been collected in the Erlenmeyer flask, lift the funnell and allow a drop of the filtrate to fall on a clean microscopic slide. Replace the funnell in the Erlenmeyer flask and allow the filtration to continue. The drop on the microscope slide will rapidly evaporate leaving behind crystals of acetylsalicylic acid. This is a qualitative test showing that the extraction of the active ingredient is successful. Report what you see on the microscope slide on your Report Sheet (3).

4. Add another 10 mL of 95 % ethanol and repeat the procedure from No.3.

5. Repeat procedure No. 4 two more times; you will use a total of 40 mL of ethanol in the four extractions. Report after each extraction if the extract carries acetylsalicylic acid. Enter these observations on your Report Sheet (4),(5) and (6).

6. When the filtration is completed and only the white moist solid is left in the filter, transfer the filter paper with its contents into a 100-mL beaker and place the beaker into a drying oven set at 110°C. Dry for 10 minutes.

7. Carefully remove the beaker from the oven. (**Caution!** The beaker is hot.) Allow it to come to room temperature. Weigh a clean and dry 25-mL

beaker on your balance. Report the weight on your Report Sheet (7). With the aid of a spatula **carefully** transfer the dried filler from the filter paper into the 25-mL beaker. Make sure that you don't spill any of the powder. Some of the dried filler may stick to the paper a bit and you may have to scrape the paper with the spatula. Weigh the 25-mL beaker with its content on your balance. Report the weight on your Report Sheet (8).

8. Test the dried filler with a drop of Hanus iodine solution. A blue coloration will signal that it contains starch. Report it on your Report Sheet (14).

Chemicals and Equipment

1. Aspirin tablets
2. Mortar and pestle (100 mL capacity)
3. 95 % ethanol
4. Filter paper (Whatman No.2, 7 cm)
5. Balance
6. Drying oven at 110°C
7. Hanus iodine solution.
8. Microscope slides
9. 100-mL beaker
10. 25-ml beaker

NAME SECTION DATE

PARTNER GRADE

Experiment 25

PRE-LAB QUESTIONS

1. Consider the statement: "Like dissolves like". Explain why acetylsalicylic acid is soluble in ethanol.

2. Consider the same statement again. Explain why starch and cellulose are not soluble in ethanol?

3. The normal adult aspirin tablet contains 5.4 grains of aspirin. If one grain is 64.8 mg, how many mg of aspirin is in one tablet?

NAME ______ SECTION ______ DATE ______

PARTNER ______ GRADE ______

Experiment 25

REPORT SHEET

1. Weight of aspirin tablets ______
2. Number of aspirin tablets in your sample ______
3. Does your first extract contain acetylsalicylic acid? ______
4. Does your second extract contain acetylsalicylic acid? ______
5. Does your third extract contain acetylsalicylic acid? ______
6. Does your fourth extract contain acetylsalicylic acid? ______
7. Weight of the empty 25-mL beaker ______
8. Weight of the 25-mL beaker and filler ______
9. Weight of the filler : (8)-(7) ______
10. Percent of filler in tablets : ((9)/(1))x100 ______
11. Weight of one tablet : (1)/(2) ______
12. Weight of filler per tablet : (11)x(10)/100 ______
13. Weight of acetylsalicylic acid per tablet : (11)-(12) ______
14. Does your filler contain starch? ______

POST-LAB QUESTIONS

1. According to your calculations does your aspirin tablet contain more, the same or less active ingredient than the average adult dosage (5.4 grains)?

2. If your ethanol extract contained a filler in addition to the active ingredient, acetylsalicylic acid, how would that effect your calculations of the dosage of the aspirin tablet?

3. If you did not dry your filler to sufficient dryness in the drying oven, how would that effect your calculations of the dosage of the aspirin tablet?

4. On the basis of the Hanus iodine test performed, what can you say about the nature of the filler in your aspirin tablet?

5. On the basis of your observations of residual crystals on the microscope slide, what can you say about the completeness of the extraction procedure?

APPENDIX 1

List of Apparatus and Equipment in Student's Locker

Amount and description

(1) Beaker, 50 mL
(1) Beaker, 100 mL
(1) Beaker, 250 mL
(1) Beaker, 400 mL
(1) Beaker, 600 mL
(1) Clamp, test tube
(1) Cylinder, graduated by 0.1 mL, 10 mL
(1) Cylinder, graduated by 1 mL, 100 mL
(1) Dropper, medicine with rubber bulb
(1) Evaporating dish
(1) Flask, Erlenmeyer, 125 mL
(1) Flask, Erlenmeyer, 250 mL
(1) Flask, Florence, 500 mL
(1) File, triangular
(1) Funnel, short stem
(1) Gauze, wire
(1) Spatula, stainless steel
(1) Sponge
(1) Striker
(6) Test tubes, approximately 15 x 150 mm
(1) Thermometer, 150°C
(1) Tongs, crucible
(1) Wash bottle, plastic
(1) Watch glass

APPENDIX 2

List of Common Equipment and Materials in the Laboratory

Each laboratory should be equipped with hoods and safety-related items such as fire extinguisher, fire blankets, safety shower, and eye wash fountain. The equipment and materials listed here for 25 2students should be made available in each laboratory.

Acid tray
Aspirators (splashgun type) on sink faucet
Balances, triple beam (or centigram) or top-loading
Barometer
Bunsen burners, with rubber tubing
Clamps, extension
Clamps, thermometer
Clamps, utility
Containers for solid chemical waste disposal
Containers for liquid organic waste disposal
Detergent for washing glass wares
Filter papers
Glass rods, 4 and 6 mm OD
Glass tubings, 6 and 8 mm OD
Glycerol (glycerine) in dropper bottles
Hot plates
Ice maker
Paper towel dispensers
Racks, test tube
Rings, support, iron, 76 mm OD
Ring stands
Rubber tubings, pressure
Rubber tubings, soft (0.25 in. OD)
Water, deionized or distilled
Weighing dishes, polystyrene, disposable, 73 x 73 x 25 mm
Weighing papers

APPENDIX 3

Special Equipment and Chemicals

In the instructions below every time a solution is to be made up in "water" you must use distilled water.

Experiment 1 Structure in organic compounds: use of molecular models

Special Equipment

(50)	Black spheres - 4 holes
(300)	Yellow spheres - 1 hole
(50)	Colored spheres (e.g. green) - 1 hole
(25)	Blue spheres - 2 holes
(400)	Sticks
(25)	Protractors
(50)	Springs (Optional)

Experiment 2 Identification of hydrocarbons

Special Equipment

(2 vials)	Litmus paper, blue
(250)	100 x 13 mm test tubes

Chemicals

(25 g)	Iron filings or powder

The following solutions should be placed in dropper bottles.

(100 mL)	Hexane
(100 mL)	Cyclohexene
(100 mL)	Toluene
(100 mL)	Ligroine (b.p. 90-110°C)
(100 mL)	1% Br_2 in hexane (wear goggles and gloves; prepare under hood): mix 1.0 mL Br_2 with enough hexane to make 100 mL
(100 mL)	1% aqueous $KMnO_4$: dissolve 1.0 gm in 50 mL distilled water by gently heating for 1 hour, cool, filter; dilute to 100 mL. Store in a dark brown dropper bottle.
(100 mL)	Concentrated H_2SO_4
(100 mL)	Unknown A = hexane
(100 mL)	Unknown B = cyclohexene
(100 mL)	Unknown C = toluene

Experiment 3 Column and paper chromatography: separation of plant pigments

Special equipment

(50)	Melting point capillaries open at both ends
(25)	25 mL burets
(1 jar)	glass wool
(25)	Filter papers (Whatman no.1), 20x10 cm
(3)	Heat lamp (optional)
(25)	Ruler with both English and metric scale
(1)	Stapler
(15)	Hot plates with or without water bath

Chemicals

(1 lb)	Tomato paste
(500 g)	Aluminum oxide (alumina)
(500 mL)	95% ethanol
(250 mL)	Dichloromethane
(200 mL)	Petroleum ether
(500 mL)	Eluting solvent: mix 450 mL petroleum ether with 10 mL toluene and 40 mL acetone.
(10 mL)	0.5% β-carotene solution: Dissolve 50 mg in 10 mL petroleum ether. Wrap the vial in aluminum foil to protect from light and keep in refrigerator until used.
(150 mL)	Saturated bromine water
(500 mg)	Iodine crystals

Experiment 4 Identification of alcohols and phenols

Special Equipment

(125)	Corks (for test tubes 100 x 13 mm)
(125)	Corks (for test tubes 150 x 18 mm)
(25)	Hot plate
(5 rolls)	Indicator paper (pH 1-12)

Chemicals

The following solutions should be placed in dropper bottles.

(100 mL)	Acetone (reagent grade)
(100 mL)	1-Butanol
(100 mL)	2-Butanol
(100 mL)	t-Butyl alcohol
(200 mL)	20% aqueous phenol: dissolve 80 g of phenol in 20 mL distilled water; dilute to 400 mL
(100 mL)	Lucas reagent (prepare under hood; wear goggles and gloves): cool 100 mL of concentrated HCl in an ice bath; with stirring, add 150 g anhydrous $ZnCl_2$ to the cold acid
(150 mL)	Chromic acid solution (prepare under hood; wear goggles and gloves): dissolve 30 g chromic oxide, CrO_3, in 30 mL concentrated H_2SO_4. Carefully add this solution to 90 mL water.
(100 mL)	2.5% ferric chloride solution: dissolve 2.5 g anhydrous $FeCl_3$ in 50 mL water; dilute to 100 mL
(100 mL)	Iodine in KI solution: mix 20 g of KI and 10 g of I_2 in 100 mL water
(250 mL)	6 M sodium hydroxide, 6 M NaOH: dissolve 60 g NaOH in 100 mL water. Dilute to 250 mL with water.
(100 mL)	Unknown A = 1-butanol
(100 mL)	Unknown B = 2-butanol
(100 mL)	Unknown C = t-butyl alcohol
(100 mL)	Unknown D = 20% aqueous phenol

Experiment 5 Identification of aldehydes and ketones

Special Equipment

(250)	Corks (to fit 100 x 13 mm test tube)
(125)	Corks (to fit 150 x 18 mm test tube)
(25)	Hot plate

Chemicals

The following solutions should be placed in dropper bottles.

(100 mL)	Acetone (reagent grade)
(100 mL)	Isovaleraldehyde
(100 mL)	Benzaldehyde (freshly distilled)
(100 mL)	Cyclohexanone
(100 mL)	Bis(2-ethoxymethyl) ether
(150 mL)	Chromic acid reagent: dissolve 30 g chromic oxide (CrO_3) in 30 mL concentrated H_2SO_4. Add carefully to 90 mL water.
	Tollens' reagent
(100 mL)	Solution A: dissolve 9.0 g silver nitrate in 90 mL of water; dilute to 100 mL
(100 mL)	Solution B: 10 g NaOH dissolved in enough water to make 100 mL
(100 mL)	10% ammonia water: 33.3 mL of 30% reagent grade ammonium hydroxide diluted to 100 mL
(200 mL)	Benedict's reagent: combine 34.6 g of hydrated sodium citrate and 20.0 g of anhydrous sodium carbonate with 160 mL water; heat to dissolve. Dissolve 3.5 g cupric sulfate pentahydrate in 10 mL water. Combine the two solutions and add enough water to bring to 200 mL.
(100 mL)	6 M sodium hydroxide, 6 M NaOH: dissolve 24 g NaOH in enough water to make 100 mL
(500 mL)	Iodine-KI solution: mix 100 g of KI and a 50 g of iodine in enough distilled water to make 500 mL

(100 mL)	2,4-Dinitrophenylhydrazine reagent: dissolve 3.0 g of 2,4-dinitrophenylhydrazine in 15 mL concentrated H_2SO_4. In a beaker, mix together 10 mL water and 75 mL 95% ethanol. With vigorous stirring slowly add the 2,4-dinitrophenylhydrazine solution to the aqueous ethanol mixture. After thorough mixing, filter by gravity through a fluted filter.
(100 mL)	95% ethanol
(100 mL)	Unknown A = isovaleraldehyde
(100 mL)	Unknown B = benzaldehyde
(100 mL)	Unknown C = cyclohexanone
(100 mL)	Unknown D = acetone

Experiment 6 Carboxylic acids and esters

Special Equipment

(5 rolls)	pH paper (range 1-12)
(100)	Disposable Pasteur pipets
(5 vials)	Litmus paper, blue
(25)	Hot plates

Chemicals

(10 g)	Salicylic acid
(10 g)	Benzoic acid

The following solutions are placed in dropper bottles.

(50 mL)	Formic acid
(75 mL)	Acetic acid
(25 mL)	Benzyl alcohol
(50 mL)	Ethyl alcohol
(25 mL)	Isobutyl alcohol
(25 mL)	Isoamyl alcohol
(50 mL)	Methyl alcohol
(25 mL)	Methyl salicylate
(200 mL)	6 M hydrochloric acid, 6 M HCl: take 300 mL concentrated HCl and bring to 500 mL
(100 mL)	3 M hydrochloric acid, 3 M HCL: take 50 mL 6 M HCl and bring to 100 mL
(300 mL)	6 M sodium hydroxide, 6 M NaOH: dissolve 72 g NaOH in enough water to bring to 300 mL
(150 mL)	2 M sodium hydroxide, 2 M NaOH: take 50 mL 6 M NaOH and bring to 150 mL
(25 mL)	Concentrated sulfuric acid, H_2SO_4

Experiment 7 Amines and amides

Special Equipment

(2 rolls)	pH paper (range 0 to 12)
(25)	Hot plates

Chemicals

(20 g)	Acetamide

The following solutions should be placed in dropper bottles.

(25 mL)	Triethylamine
(25 mL)	Aniline
(25 mL)	N,N-Dimethylaniline
(100 mL)	Diethyl ether (ether)
(100 mL)	6 M ammonia solution, 6 M NH_3: 40 mL concentrated NH_4OH diluted to 100 mL with water
(100 mL)	6 M hydrochloric acid, 6 M HCl: 50 mL concentrated HCl diluted to 100 mL with water
(50 mL)	Concentrated hydrochloric acid, HCl
(250 mL)	6 M sulfuric acid, 6 M H_2SO_4: pour 82.5 mL concentrated H_2SO_4 into 125 mL cold water. Stir slowly. Dilute to 250 mL with water.
(250 mL)	6 M sodium hydroxide, 6 M NaOH: dissolve 60 g NaOH in 100 mL water. Dilute to 250 mL with water.

Experiment 8 Polymerization reactions

Special equipment

(25)	Hot plates
(25)	Cylindrical paper rolls or sticks
(25)	Bent wire approximatly 10-cm long
(25)	10-mL pipets or syringes
(25)	Spectroline pipet fillers
(25)	Beaker tongs

Chemicals

(750 mL)	Styrene
(250 mL)	Xylene
(75 mL)	Tertiary butyl peroxide benzoate (also called tertiary butyl benzoyl peroxide). Store at 4°C.
(75 mL)	20% NaOH: dissolve 15 g NaOH in 75 mL water
(300 mL)	5% adipoyl chloride: dissolve 15 g adipoyl chloride in 300 mL cyclohexane
(300 mL)	5% hexamethylene diamine: dissolve 15 g hexamethylene diamine in 300 mL water
(200 mL)	80% formic acid: add 40 mL water to 160 mL formic acid

Experiment 9 Preparation of aspirin

Special Equipment

(25)	Büchner funnel (85-mm OD)
(25)	Filtervac or neoprene adapter
(1 box)	Filter paper (7.0 cm, Whatman no. 2)
(25)	500-mL filter flask
(25)	Hot Plate

Chemicals

(1 bottle)	Boiling chips
(200 g)	Salicylic acid
(25)	Commercial aspirin tablets
(100 mL)	Concentrated sulfuric acid, H_2SO_4 (in a dropper bottle)
(100 mL)	1% ferric chloride: dissolve 1 g $FeCl_3 \cdot 6\ H_2O$ in enough distilled water to make 100 mL (in a dropper bottle)
(200 mL)	Acetic anhydride, freshly opened bottle
(500 mL)	95% ethanol

Experiment 10 Isolation of caffeine from tea leaves

Special Equipment

(25)	Hot plates
(25)	125-mL separatory funnels
(1 box)	Filter paper; 7.0 cm, fast flow (Whatman no. 1)
(1 vial)	Melting point capillaries
(25)	Small sample vials
(1)	Stapler

Chemicals

(50)	Tea bags
(1 bottle)	Boiling chips
(25 g)	Sodium sulfate, anhydrous, Na_2SO_4
(50 g)	Sodium carbonate, anhydrous, Na_2CO_3
(500 mL)	Methylene chloride, CH_2Cl_2

Experiment 11 Carbohydrates

Special Equipment

(50)	Medicine droppers
(125)	Microtest tubes or 25 white spot plates
(2 rolls)	Litmus paper, red

Chemicals

(20 g)	Boiling chips
(400 mL)	Fehling's reagent (solution A and B, from Fisher Scientific Co.)
(200 mL)	3 M NaOH: dissolve 24 g NaOH in water and bring it to 200 mL volume
(200 mL)	2% starch solution: place 4 g soluble starch in a beaker. With vigorous stirring, add 10 mL water to form a thin paste. Boil 190 mL water in another beaker. Add the starch paste to the boiling water and stir until the solution becomes clear. (Store in a dropper bottle).
(200 mL)	2% sucrose: dissolve 4 g sucrose in 200 mL water
(200 mL)	3 M sulfuric acid: add 33 mL concentrated H_2SO_4 to 150 mL ice cold water; pour the sulfuric acid slowly along the walls of the beaker, this way it will settle on the bottom without much mixing; stir slowly in order not to generate too much heat; when fully mixed bring the volume to 200 mL.
(100 mL)	2% fructose: dissolve 2 g fructose in 100 mL water. (Store in a dropper bottle)
(100 mL)	2% glucose: dissolve 2 g glucose in 100 mL water. (Store in a dropper bottle)
(100 mL)	2% lactose: dissolve 2 g lactose in 100 mL water. (Store in a dropper bottle)
(100 mL)	0.01 M iodine in KI: dissolve 1.2 g KI in 80 mL water. Add 0.25 g I_2. Stir until the iodine dissolves. Dilute the solution to 100 mL volume. (Store in a dark dropper bottle)

Experiment 12 Preparation and properties of a soap

Special Equipment

(25)	Büchner funnels (85 mm OD)
(25)	No. 7 one-hole rubber stoppers
(1 box)	Filter paper (Whatman no.2) 7.0 cm
(1 roll)	pHydrion paper (pH range 0 to 12)

Chemicals

(20 g)	Boiling chips
(1 L)	95% ethanol
(1 L)	Saturated sodium chloride: dissolve 360 g NaCl in 1 L water
(1 L)	25% NaOH: dissolve 250 g NaOH in 1 L water.
(1 L)	Vegetable oil
(100 mL)	5% iron (III) chloride: dissolve 5 g $FeCl_3 \cdot 6H_2O$ in 100 mL water. (Store in a dropper bottle)
(100 mL)	5% $CaCl_2$: dissolve 5 g $CaCl_2 \cdot H_2O$ in 100 mL water. (Store in a dropper bottle)
(100 mL)	Mineral oil (Store in a dropper bottle)
(100 mL)	5% $MgCl_2$: dissolve 5 g $MgCl_2$ in 100 mL water. (Store in a dropper bottle)

Experiment 13 Preparation of Hand Cream

Special Equipment

(25) Bunsen burners

Chemicals

(100 mL)	Triethanolamine
(40 mL)	Propylene glycol (1,2 Propanediol)
(500 g)	Stearic acid
(40 g)	Methyl stearate
(400 g)	Lanolin
(400 g)	Mineral oil

Experiment 14 Isolation of Lipids from Egg Yolk

Special Equipment

(12) Hard boiled eggs
(13) Steam baths
(13) Hot plates
(2) Waste jars

Chemicals

(4 L) Acetone
(1.5 L) Ethyl ether

Experiment 15 Analysis of Lipids

Special Equipment

(75) Capillary melting point tubes
(5) Melting point apparatus
(25) Hot plates

Chemicals

(250 mL) Molybdate solution: dissolve 0.8 g $(NH_4)_6Mo_7O_{24}\cdot 4H_2O$ in 30 mL water. Put in an ice bath. Pour slowly 20 mL concentrated sulfuric acid into the solution and stir it slowly. After cooling to room temperature bring the volume to 250 mL.
(50 mL) 0.1 M ascorbic acid solution: dissolve 0.88 g ascorbic acid (vitamin C) in water and bring to 50 mL volume This must be prepared fresh every week and stored at 4°C.
(250 mL) 6 M NaOH: dissolve 60 g NaOH in water and bring the volume to 250 mL
(250 mL) 6 M HNO_3: pipet 63 mL concentrated HNO_3 into a 250 mL volumetric flask and bring to volume with water
(200 mL) Chloroform
(75 mL) Acetic anhydride
(50 mL) Concentrated H_2SO_4
(75 g) $KHSO_4$

Experiment 16 TLC separation of amino acids

Special Equipment

(1)	Drying oven, 105-110°C
(2)	Heat lamps or hair dryers
(50)	15 x 6.5 cm silica gel TLC plates
(25)	Rulers, metric scale
(25)	Polyethylene, surgical gloves
(150)	Capillary tubes, open on both ends
(1 roll)	Aluminum foil

Chemicals

(25 mL)	0.12% aspartic acid solution: dissolve 30 mg aspartic acid in 25 mL distilled water
(25 mL)	0.12% phenylalanine solution: dissolve 30 mg phenylalanine in 25 mL distilled water
(25 mL)	0.12% leucine solution: dissolve 30 mg leucine in 25 mL distilled water
(25 mL)	Aspartame solution: dissolve 150 mg Equal sweetener powder in 25 mL distilled water
(50 mL)	3 M HCl solution: place 10 mL distilled water into a 50 mL volumetric flask. Add slowly 12.46 mL of concentrated HCl and bring it to volume with distilled water.
(1 L)	Solvent mixture: mix 600 mL n-butanol with 150 mL acetic acid and 240 mL distilled water
(1 can)	Ninhydrin spray reagent.(0.2% ninhydrin in ethanol or acetone).Do not use any reagent older than 6 months old.
(1 can)	Diet Coca-Cola
(4 packets)	Equal, NutraSweet, sweeteners

Experiment 17 Acid-base properties of amino acids

Special Equipment

(10)	pH meters or
(5 rolls)	pHydrion short-range papers from each range: pH: 0.0 to 3.0; 3.0 to 5.5; 5.2 to 6.6; 6.0 to 8.0; 8.0 to 9.5 and 9.0 to 12.0
(25)	20-mL pipets
(25)	50-mL burets
(25)	Spectroline pipet fillers
(25)	Pasteur pipets

Chemicals

(500 mL)	0.25 M NaOH: dissolve 5 g NaOH in water and bring to 500 mL volume
(750 mL)	0.1 M alanine solution: dissolve 6.68 g L-alanine in 500 mL water; add sufficient 1 M HCl to bring the pH to 1.5. Bring the volume to 750 mL with water. or do as above but use either 5.63 g glycine or 9.84 g leucine or 12.39 g phenylalanine or 8.79 g valine.

Experiment 18 Isolation and identification of casein

Special Equipment

(25)	Hot plates
(25)	600-mL beakers
(25)	Büchner funnels (O.D. 85 mm) in No. 7 hole rubber stopper
(7 boxes)	Whatman No.2 filter paper,7 cm
(25)	Rubber bands
(25)	Cheese cloths (6x6 in.)

Chemicals

(25 g)	Boiling chips
(1 L)	95% ethanol
(1 L)	Diethyl ether-ethanol mixture (1:1)
(0.5 gal)	Regular milk
(500 mL)	Glacial acetic acid

The following solutions should be placed in dropper bottles:

(100 mL)	Concentrated nitric acid, HNO_3,(69%)
(100 mL)	2% albumin suspension: dissolve 2 g albumin in 100 mL water
(100 mL)	2% gelatin: dissolve 2 g gelatin in 100 mL water
(100 mL)	2% glycine: dissolve 2 g glycine in 100 mL water
(100 mL)	5% copper (II) sulfate: dissolve 5 g $CuSO_4$ (or 7.85 g $CuSO_4.5H_2O$ in 100 mL water)
(100 mL)	5% lead (II) nitrate: dissolve 5 g $Pb(NO_3)_2$ in 100 mL water
(100 mL)	5% mercury (II) nitrate: dissolve 5 g $Hg(NO_3)_2$ in 100 mL water
(100 mL)	Ninhydrin reagent: dissolve 3 g ninhydrin in 100 mL acetone
(100 mL)	10% NaOH: dissolve 10 g NaOH in 100 mL water
(100 mL)	1% tyrosine: dissolve 1 g tyrosine in 100 mL water
(100 mL)	5% $NaNO_3$: dissolve 5 g $NaNO_3$ in 100 mL water

Experiment 19 Isolation and identification of DNA from yeast

Special equipment

(12)	Mortars
(12)	Pestles
(6)	Desk top clinical centrifuges (swinging bucket rotor) (optional)

Chemicals

(100 g)	Baker's yeast, freshly purchased
(500 g)	Acid washed sand
(1 L)	Saline-CTAB isolation buffer. Dissolve 20 g hexadecyltrimethylammonium bromide (CTAB, Sigma 45882) , 2 mL 2-mercaptoethanol, 7.44 g ethylenediamine tetraacetate (EDTA, Sigma ED2SS), 8.77 g NaCl in 1 L Tris buffer. The Tris buffer is prepared by dissolving 12.1 g Tris in 700 mL water; adjusting the pH to 8 by titrating with 4 M HCl. Bring the volume to 1 L.
(200 mL)	6 M $NaClO_4$ solution. Dissolve 147 g $NaClO_4$ in water and bring the volume to 200 mL.
(100 mL)	Citrate buffer. Dissolve 0.88 g NaCl and 0.39 g sodium citrate in 100 mL water.
(1 L)	Chloroform-isoamyl alcohol mixture. To 960 mL chloroform add 40 mL isoamyl alcohol. Mix throughly.
(2 L)	Isopropyl alcohol
(50 mL)	1% glucose solution. Dissolve 0.5 g D-glucose in 50 mL water.
(50 mL)	1% ribose solution. Dissolve 0.5 g D-ribose in 50 mL water.
(50 mL)	1% deoxyribose solution. Dissolve 0.5 g 2-deoxy-D-ribose in 50 mL water.
(200 mL)	95% ethanol.
(500 mL)	Diphenylamine reagent. This must be prepared shortly before lab use. Dissolve 7.5 g reagent grade diphenylamine (Sigma D3409) in 50 mL glacial acetic acid. Add 7.5 mL concentrated sulfuric acid. Prior to use add 2.5 mL 1.6% acetaldehyde (made by dissolving 0.16 g acetaldehyde in 10 mL water).

Experiment 20 Kinetics of urease catalyzed decomposition of urea

Special Equipment

(25)	5-mL pipets
(25)	10-mL graduated pipets
(25)	10-mL volumetric pipets
(25)	50-mL burets
(25)	Buret holders
(25)	Spectroline pipet fillers

Chemicals

(3.5 L)	0.05 M Tris buffer: dissolve 21.05 g Tris buffer in water (3 L). Adjust the pH to 7.2 with 1 M HCl solution; add sufficient water to make 3.5 L. Parts of buffer solution will be used to make urea and enzyme solutions.
(2.5 L)	0.3 M urea solution: dissolve 45 g urea in 2.5 L Tris buffer
(50 mL)	1×10^{-3} M phenylmercuric acetate: dissolve 16.5 mg phenylmercuric acetate in 50 mL distilled water

CAUTION! Phenylmercuric acetate is a poison. Do not touch the chemical with hand. Do not swallow the solution.

(50 mL)	1% $HgCl_2$ solution: dissolve 0.5 g $HgCl_2$ in 100 mL water
(100 mL)	0.04% methyl red indicator: dissolve 40 mg methyl red in 100 mL water
(1.0 L)	Urease solution: prepare enzyme solution on the week of experiment and store at 4°C. Take 200 mg urease, dissolve in 1 L Tris buffer. One can buy urease with 5 to 6 units activity per mg enzyme, for example, from Nutritional Biochemical, Cleveland, Ohio.
(500 mL)	0.05 N HCl: dilute 2.5 mL concentrated HCl with water to 500 mL volume

Experiment 21 Isocitrate dehydrogenase

Special Eqipment

(15)	Büchner funnel (85 mm OD) in no. 7 one-hole rubber stopper.
(15)	200 mL filter flasks
(1 box)	Filter papers, Whatman no.1

Chemicals

(30 g)	Fresh Baker's yeast
(100 g)	Acid washed sand
(200 mL)	0.1 m $NaHCO_3$ buffer: weigh 1.68 g sodium bicarbonate add distilled water to bring the volume to 200 mL
(40 mL)	Phosphate buffer at pH 7.0: Mix 25 mL of 0.1 M KH_2PO_4 and 15 mL of 0.1 M NaOH solutions.To prepare 0.1 M NaOH weigh 0.2 g NaOH dissolve it in 20 mL distilled water and transfer it to a 50 mL volumetric flask and bring it to volume. To prepare 0.1 M KH_2PO_4 weigh 0.68 g of potassium dihydrogen phosphate add sufficient distilled water to dissolve it and bring it to 50 mL volume.
(20 mL)	0.1 M $MgCl_2$ solution: Dissolve 0.19 g magnesium chloride in 20 mL distilled water.
(20 mL)	2.5 mM ADP solution: Dissolve 23 mg ADP in 20 mL distilled water.
(20 mL)	6.0 mM NAD^+ solution: Dissolve 72 mg NAD^+ in 20 mL distilled water.
(50 mL)	10 mM sodium isocitrate solution: Dissolve 107 mg sodium isocitrate in 50 mL distilled water.
(50 g)	Celite filter aid

Experiment 22 Quantitative analysis of vitamin C contained in foods

Special Equipment

(25)	50-mL burets
(25)	Buret clamps
(25)	Ring stands
(25)	Spectroline pipet fillers
(25)	10-mL volumetric pipets
(1 box)	Cotton

Chemicals

(500 g)	Celite, filter aid
(1 can)	Hi-C orange drink
(1 can)	Hi-C grapefruit drink
(1 can)	Hi-C apple drink
(2 L)	0.01 M iodine: dissolve 5 g KI in 300 mL water. Add 2.5 g I_2, stir until dissolves. Dilute the solution with water to 2 L volume. Store in a dark bottle. **Caution!** Iodine is poisonous if taken internally.
(100 mL)	3 M HCl: dilute 25 mL concentrated HCl (36%) with water to 100 mL volume
(100 mL)	2 % starch solution: place 2 g soluble starch in a 50-mL beaker. Add 10 mL water and stir vigorously to form a paste. Boil 90 mL water in a second beaker. Add the starch paste to the boiling water. Stir until the solution becomes translucent. Cool to room temperature.

Experiment 23 Urine analysis

Special Equipment

(12)	Hydrometers (urinometers) from 1.00 to 1.40 specific gravity
(3 bottles)	Clinistix (50 reagent strips/bottle)
(2 bottles)	Urobilistix (50/reagent strips/bottle)
(2 bottles)	Phenistix (50 reagent strips/bottle)
(1 bottle)	Albustix (100 reagent strips/bottle)
(1 bottle)	Ketostix (100 reagent strip/bottle)

All these are obtainable from Ames Co., Division Miles Laboratories Inc., Elkhart, Indiana,46515. Instead of the individual "Stix"-es you may purchase 4 bottles of multipurpose Labstix (100 reagent strips/bottle).

Chemicals

(500 mL)	Normal urine. This and all other urine samples must be kept at 4°C until 30 minutes prior to the lab period. Alternatively, you may ask each student to provide fresh urine samples for analysis.
(500 mL)	"Pathological urine A": add 4 g glucose, 2 mL acetone and 2 g citric acid to 500 mL water
(500 mL)	"Pathological urine B": add 50 mg phenylpyruvate, 500 mg creatinine and 500 mg sodium phosphate, Na_3PO_4, to 500 mL water
(250 mL)	1 % glucose: dissolve 2.5 g glucose in 250 mL water
(200 mL)	0.25 % glucose: dilute 50 mL of 1 % glucose with water to 200 mL volume
(100 mL)	1 % creatinine: dissolve 1 g creatinine in 100 mL water.
(50 mL)	0.5 % creatinine:dilute 25 mL of 1 % creatinine with water to 50 mL volume
(50 mL)	0.1 % creatinine: dilute 5 mL of 1 % creatinine with water to 50 mL volume
(50 mL)	Saturated picric acid solution: dissolve 1.5 g picric acid in 50 mL warm (70°C) water. Cool to room temperature. **Caution!** Picric acid is explosive. Store only minimal quantities in the stockroom.
(100 mL)	3 M NaOH: dissolve 12 g NaOH in 100 mL water

Experiment 24 Measurement of sulfur dioxide preservative in foods

Special Equipment

(1)	Blender
(175)	100-mL volumetric flasks
(25)	10-mL pipets
(100)	1-mL graduated pipets
(75)	10-mL graduated pipets
(5)	Spectrophotometers

Chemicals

(10 g)	Raisins
(200 mL)	0.5 N NaOH: Dissolve 4.0 g NaOH in 50 mL water and bring it to 200 mL volume.
(200 mL)	0.5 N H_2SO_4 solution: Place 100 mL water in a beaker. Cool it in an ice bath. Add slowly from a graduate cylinder 10.7 mL concentrated sulfuric acid. Make sure you pour the concentrated acid slowly along the walls of the beaker. If you add it too fast the acid may splash and create severe burns. It is advisable to wear goggles and gloves during this process. Wait a few minutes. Slowly stir the solution with a glass rod and add sufficient water to bring it to 200 mL.
(1.5 L)	0.015% formaldehyde solution: Take 0.56 mL of 40% formaldehyde solution and add it to 1.5 L water.
(1 L)	Rosaniline reagent: Place 100 mg p-rosaniline.HCl (Allied Chem. Corp.) and 200 mL distilled water in a 1 L volumetric flask. Take 80 mL concentrated HCl and add 80 mL water. Stir. Add the HCl solution to the 1 L volumetric flask, mix and bring it to volume. The rosaniline reagent must stand at least 12 hrs before use.
(1.5 L)	Mercurate reagent: Use polyethylene gloves to protect your skin from touching mercurate reagent. Mercury compounds are toxic and if spills occur you should wash them immediately with copious amounts of water. Dissolve 17.6 g NaCl and 40.7 g $HgCl_2$ in 1 L water and dilute it to 1.5 L.
(500 mL)	Standard sulfur dioxide stock solution: Dissolve 85 mg $NaHSO_3$ in 500 mL water.

Experiment 25 Measurement of the active ingredient in aspirin pills

Special Equipment

(1)	Drying oven at 110°C
(25)	Mortars 100 mL capacity
(25)	Pestles
(1 box)	Filter paper, 7.0 cm diameter, Whatman #2
(1 box)	Microscope slides, 3x1 inch plain
(25)	25-mL beakers

Chemicals

(1.5 L)	95 % ethanol
(300 g)	Commercial asprin tablets
(100 mL)	Hanus iodine solution: dissolve 1.2 g KI in 80 mL water. Add 0.25 g I_2. Stir until the iodine dissolves. Add water to make 100 mL volume. (Store in dark dropper bottle)